Contents

	PAGE
Introduction to amendments 2004	5
Use of guidance	6
The Approved Documents	6
Limitation on requirements	6
Materials and workmanship	6
The Workplace (Health, Safety and Welfare) Regulations 1992	7
The Requirements	8
Section 0: Performance	12
Performance standards	12
Section 1: Pre-completion testing	16
Introduction	16
Grouping	16
Sub-grouping for new buildings	16
Sub-grouping for material change of use	17
Sets of tests in dwelling-houses (including bungalows)	17
Sets of tests in flats with separating floors but without separating walls	17
Sets of tests in flats with a separating floor and a separating wall	17
Types of rooms for testing	17
Sets of tests in rooms for residential purposes	17
Properties sold before fitting out	17
Normal programme of testing	17
Action following a failed set of tests	18
Remedial treatment	18
Material change of use	18
Approved manner of recording pre-completion testing results	18
Section 2: Separating walls and associated flanking constructions for new buildings	20
Introduction	20
Junctions between separating walls and other building elements	20
Mass per unit area of walls	21
Plasterboard linings on separating and external masonry walls	21
Cavity widths in separating cavity masonry walls	21
Wall ties in separating and external cavity masonry walls	21
Corridor walls and doors	22
Refuse chutes	22
Wall type 1: solid masonry	22
Junction requirements for wall type 1	24
Wall type 2: cavity masonry	28
Junction requirements for wall type 2	30
Wall type 3: masonry between independent panels	33
Junction requirements for wall type 3	36
Wall type 4: framed walls with absorbent material	39
Junction requirements for wall type 4	40
Section 3: Separating floors and associated flanking constructions for new buildings	42
Introduction	42
Junctions between separating floors and other building elements	42
Beam and block floors	43
Mass per unit area of floors	43
Ceiling treatments	43
Floor type 1: concrete base with ceiling and soft floor covering	44
Junction requirements for floor type 1	45
Floor type 2: concrete base with ceiling and floating floor	48
Floating floors (floating layers and resilient layers)	49
Junction requirements for floor type 2	51
Floor type 3: timber frame base with ceiling and platform floor	53
Junction requirements for floor type 3	54
Section 4: Dwelling-houses and flats formed by material change of use	57
Introduction	57
Work to existing construction	58
Corridor walls and doors	58
Wall treatment 1: independent panel(s) with absorbent material	59
Floor treatment 1: independent ceiling with absorbent material	60
Floor treatment 2: platform floor with absorbent material	61
Stair treatment: stair covering an independent ceiling with absorbent material	62
Junction requirements for material change of use	62

PAGE

Section 5: Internal walls and floors for new buildings 64
Introduction 64
Doors 64
Layout 64
Junction requirements for internal walls 64
Junction requirements for internal floors 64
Internal wall type A: Timber or metal frames with plasterboard linings on each side of frame 64
Internal wall type B: Timber or metal frames with plasterboard linings on each side of frame and absorbent material 65
Internal wall type C: Concrete block wall, plaster or plasterboard finish on both sides 65
Internal wall type D: Aircrete block wall plaster or plasterboard finish on both sides 65
Internal floor type A: Concrete planks 66
Internal floor type B: Concrete beams with infilling blocks, bonded screed and ceiling 66
Internal floor type C: Timber or metal joist, with wood based board and plasterboard ceiling, and absorbent material 66

Section 6: Rooms for residential purposes 67
Introduction 67
Separating walls in new buildings containing rooms for residential purposes 67
Corridor walls and doors 67
Separating floors in new buildings containing rooms for residential purposes 67
Rooms for residential purposes resulting from a material change of use 67
Junction details 68
Room layout and building services design considerations 68

Section 7: Reverberation in the common internal parts of buildings containing flats or rooms for residential purposes 69
Introduction 69
Method A 69
Method B 69
Report format 70
Worked example 70

Section 8: Acoustic conditions in schools 72

Annex A: Method for calculating mass per unit area 73
A1 Wall mass 73
A2 Formula for calculation of wall leaf mass per unit area 73
A3 Simplified equations 73
A4 Mass per unit area of surface finishes 74
A5 Mass per unit area of floors 74

Annex B: Procedures for sound insulation testing 75
B1 Introduction 75
B2 Field measurement of sound insulation of separating walls and floors for the purposes of Regulation 40 and Regulation 20(1) & (5) 75
B3 Laboratory measurements 76
B4 Information to be included in test reports 77

Annex C: Glossary 78

Annex D: References 80
D1 Standards 80
D2 Guidance 80
D3 Legislation 81

Annex E: Design details approved by Robust Details Ltd 82

DIAGRAMS
0.1 Requirement E1 14
0.2 Requirement E2(a) 15
0.3 Requirement E2(b) 15
2.1 Types of separating wall 20
2.2 Wall type 1.1 23
2.3 Wall type 1.2 23
2.4 Wall type 1.3 24
2.5 Wall type 1 – external cavity wall with masonry inner leaf 24
2.6 Wall type 1 – bonded junction – masonry inner leaf of external cavity wall with solid separating wall 25
2.7 Wall type 1 – tied junction – external cavity wall with internal masonry wall 25
2.8 Wall type 1 – position of openings in masonry inner leaf of external cavity wall 25
2.9 Wall type 1 – external cavity wall with timber frame inner leaf 26
2.10 Wall type 1 – internal timber floor 26
2.11 Wall type 1 – internal concrete floor 26
2.12 Wall type 1 – concrete ground floor 27
2.13 Wall type 1 – ceiling and roof junction 27
2.14 External cavity wall at eaves level 27
2.15 Wall type 2.1 28

		PAGE
2.16	Wall type 2.2	29
2.17	Wall type 2.3	29
2.18	Wall type 2.4	30
2.19	Wall types 2.1 and 2.2 – external cavity wall with masonry inner leaf	30
2.20	Wall types 2.3 and 2.4 – external cavity wall with masonry inner leaf – stagger	31
2.21	Wall type 2 – tied junction – external cavity wall with internal masonry wall	31
2.22	Wall type 2 – external cavity wall with timber frame inner leaf	31
2.23	Wall type 2 – internal timber floor	32
2.24	Wall type 2 – internal concrete floor and concrete ground floor	32
2.25	Wall type 2 – ceiling and roof junction	32
2.26	External cavity wall at eaves level	33
2.27	Wall type 3.1 with independent composite panels	34
2.28	Wall type 3.1 with independent plasterboard panels	34
2.29	Wall type 3.2 with independent composite panels	35
2.30	Wall type 3.3 with independent composite panels	35
2.31	Wall type 3 – external cavity wall with masonry inner leaf	36
2.32	Wall type 3 – external cavity wall with internal timber wall	36
2.33	Wall type 3 – internal timber floor	37
2.34	Wall types 3.1 and 3.2 – internal concrete floor	37
2.35	Wall types 3.1 and 3.2 – ceiling and roof junction	38
2.36	External cavity wall at eaves level	38
2.37	Wall type 4.1	40
2.38	Wall type 4 – external cavity wall with timber frame inner leaf	40
3.1	Types of separating floor	42
3.2	Ceiling treatments A, B and C	44
3.3	Floor type 1.1C – floor type 1.1 with ceiling treatment C	45
3.4	Floor type 1.2B – floor type 1.2 with ceiling treatment B	45
3.5	Floor type 1.2B – external cavity wall with masonry inner leaf	46
3.6	Floor type 1 – floor penetrations	46
3.7	Floor type 1.1C – wall type 1	47
3.8	Floor type 1.2B – wall type 1	47

		PAGE
3.9	Floor types 1.1C and 1.2B – wall type 2	47
3.10	Floor type 1.1C – wall types 3.1 and 3.2	48
3.11	Floating floors (a) and (b)	49
3.12	Floor type 2.1C(a) – floor type 2.1 with ceiling treatment C and floating floor (a)	50
3.13	Floor type 2.1C(b) – floor type 2.1 with ceiling treatment C and floating floor (b)	50
3.14	Floor type 2.2B(a) – floor type 2.2 with ceiling treatment B and floating floor (a)	50
3.15	Floor type 2.2B(b) – floor type 2.2 with ceiling treatment B and floating floor (b)	50
3.16	Floor type 2 – external cavity wall with masonry internal leaf	51
3.17	Floor type 2 – floor penetrations	52
3.18	Floor types 2.2B(a) and 2.2B(b) – wall type 1	52
3.19	Floor type 2.1C – wall type 3.1 and 3.2	52
3.20	Floor type 3.1A	53
3.21	Floor type 3 – floor penetrations	55
3.22	Floor type 3 – wall type 1	55
3.23	Floor type 3 – wall type 2	56
4.1	Treatments for material change of use	58
4.2	Wall treatment 1	59
4.3	Floor treatment 1	60
4.4	Floor treatment 1 – high window head detail	60
4.5	Floor treatment 1 – wall treatment 1	61
4.6	Floor treatment 2	61
4.7	Floor treatment 2 – wall treatment 1	62
4.8	Stair treatment	62
4.9	Floor penetrations	63
5.1	Internal wall type A	65
5.2	Internal wall type B	65
5.3	Internal wall type C	65
5.4	Internal wall type D	66
5.5	Internal floor type A	66
5.6	Internal floor type B	66
5.7	Internal floor type C	66
6.1	Ceiling void and roof space (only applicable to rooms for residential purposes)	68
A.1	Block and mortar dimensions	73
A.2	Beam and block floor dimensions	74

PAGE

TABLES

0.1a Dwelling-houses and flats – performance standards for separating walls, separating floors, and stairs that have a separating function 12

0.1b Rooms for residential purposes – performance standards for separating walls, separating floors, and stairs that have a separating function 12

0.2 Laboratory values for new internal walls and floors within dwelling-houses, flats and rooms for residential purposes, whether purpose built or formed by material change of use 13

2.1 Separating wall junctions reference table 21

3.1 Separating floor junctions reference table 43

7.1 Absorption coefficient data for common materials in buildings 70

7.2 Example calculation for an entrance hall (Method B) 71

A.1 Blocks laid flat 73

A.2 Blocks laid on edge 74

Introduction to amendments 2004*

The current edition of Part E in Schedule 1 to the Building Regulations 2000 (as amended) came into force on 1 July 2003. At the same time a new Regulation 20A was introduced into the Building Regulations 2000, and a new Regulation 12A was introduced into the Building (Approved Inspectors, etc.) Regulations 2000. Regulations 20A and 12A introduced pre-completion testing for sound insulation as a means of demonstrating compliance. Pre-completion testing has applied to rooms for residential purposes, houses and flats formed by conversion of other buildings since 1 July 2003, and it will apply to new houses and flats from 1 July 2004. Also, from 1 July 2004, use of robust details in new houses and flats will be accepted as an alternative to testing.

Robust details are high performance separating wall and floor constructions (with associated construction details) that are expected to be sufficiently reliable not to need the check provided by pre-completion testing.

The introduction of robust details has necessitated the amendment of Regulations 20A and 12A. The amendments have been made by the Building (Amendment) Regulations 2004 and the Building (Approved Inspectors, etc.) (Amendment) Regulations 2004. Regulations 20A and 12A are reproduced in Approved Document E, 2003 Edition; and so amendments to that Approved Document are needed to pick up the changes.

Section 0 of Approved Document E, 2003 edition, has also been amended to explain the use of robust details.

The 2003 edition of Part E introduced a new class of dwelling known as a *room for residential purposes*, which covers hostel types of accommodation and hotel rooms. The expression 'room for residential purposes' is defined in Regulation 2 of the Building Regulations 2000 and the definition is reproduced in Approved Document E, 2003 Edition. However, the definition has been interpreted in different ways by building control bodies, particularly in respect of student halls of residence, and it has, therefore, been clarified, by means of the Building (Amendment) Regulations 2004.

A number of errors have been found in Approved Document E, 2003 Edition, and also some guidance that is unclear.

This Amendment document sets out the text of the amended regulations 20A and 12A, the clarified definition of *room for residential purposes*, and also amendments, corrections and clarifications to the text of Approved Document E, 2003 Edition. This document is approved by the Secretary of State from 1 July 2004.

Buildings Division
Office of the Deputy Prime Minister
June 2004

*On this page, references to the 2000 Regulations have not been updated to reflect changes in the 2010 Regulations.

Use of guidance

THE APPROVED DOCUMENTS

This document is one of a series that has been approved and issued by the Secretary of State for the purpose of providing practical guidance with respect to the requirements of Schedule 1 to, and Regulation 7 of, the Building Regulations 2010 (SI 2010/2214) for England and Wales.

At the back of this document is a list of all the documents that have been approved and issued by the Secretary of State for this purpose.

Approved Documents are intended to provide guidance for some of the more common building situations. However, there may well be alternative ways of achieving compliance with the requirements. Thus there is no obligation to adopt any particular solution contained in an Approved Document if you prefer to meet the relevant requirement in some other way.

Other requirements

The guidance contained in an Approved Document relates only to the particular requirements of the Regulations which the document addresses. The building work will also have to comply with the requirements of any other relevant paragraphs in Schedule 1 to the Regulations.

There are Approved Documents which give guidance on each of the parts of Schedule 1 and on Regulation 7.

LIMITATION ON REQUIREMENTS

In accordance with Regulation 8, the requirements in Parts A to D, F to K and N (except for paragraphs H2 and J7) of Schedule 1 to the Building Regulations do not require anything to be done except for the purpose of securing reasonable standards of health and safety for persons in or about buildings (and any others who may be affected by buildings or matters connected with buildings). This is one of the categories of purpose for which building regulations may be made.

Paragraphs H2 and J7 are excluded from Regulation 8 because they deal directly with prevention of the contamination of water. Parts E and M (which deal, respectively, with resistance to the passage of sound, and access and facilities for disabled people) are excluded from Regulation 8 because they address the welfare and convenience of building users. Part L is excluded from Regulation 8 because it addresses the conservation of fuel and power. All these matters are amongst the purposes, other than health and safety, that may be addressed by Building Regulations.

MATERIALS AND WORKMANSHIP

Materials and workmanship

Any building work which is subject to the requirements imposed by Schedule 1 to the Building Regulations shall be carried out in accordance with regulation 7. Guidance on meeting these requirements on materials and workmanship is contained in Approved Document 7.

Building Regulations are made for specific purposes, primarily the health and safety, welfare and convenience of people and for energy conservation. Standards and other technical specifications may provide relevant guidance to the extent that they relate to these considerations. However, they may also address other aspects of performance or matters which, although they relate to health and safety etc., are not covered by the Building Regulations.

When an Approved Document makes reference to a named standard, the relevant version of the standard to which it refers is the one listed at the end of the publication. However, if this version has been revised or updated by the issuing standards body, the new version may be used as a source of guidance provided it continues to address the relevant requirements of the Regulations.

THE WORKPLACE (HEALTH, SAFETY AND WELFARE) REGULATIONS 1992

The Workplace (Health, Safety and Welfare) Regulations 1992 contain some requirements which affect building design. The main requirements are now covered by the Building Regulations, but for further information see – *Workplace health, safety and welfare.* L24 *Workplace (Health, Safety and Welfare) Regulations 1992. Approved Code of Practice and Guidance,* 1998. ISBN 0 71760 413 6.

The Workplace (Health, Safety and Welfare) Regulations 1992 apply to the common parts of flats and similar buildings if people such as cleaners and caretakers are employed to work in these common parts. Where the requirements of the Building Regulations that are covered by this Part do not apply to dwellings, the provisions may still be required in the situations described above in order to satisfy the Workplace Regulations.

The Requirements

This Approved Document, which took effect on 1 July 2003, deals with the Requirements of Part E of Schedule 1 to the Building Regulations 2010.

Requirement	*Limits on application*
Protection against sound from other parts of the building and adjoining buildings	
E1. Dwelling-houses, flats and rooms for residential purposes shall be designed and constructed in such a way that they provide reasonable resistance to sound from other parts of the same building and from adjoining buildings.	
Protection against sound within a dwelling-house etc.	
E2. Dwelling-houses, flats and rooms for residential purposes shall be designed and constructed in such a way that: (a) internal walls between a bedroom or a room containing a water closet, and other rooms; and (b) internal floors provide reasonable resistance to sound.	Requirement E2 does not apply to: (a) an internal wall which contains a door; (b) an internal wall which separates an en suite toilet from the associated bedroom; (c) existing walls and floors in a building which is subject to a material change of use.
Reverberation in the common internal parts of buildings containing flats or rooms for residential purposes	
E3. The common internal parts of buildings which contain flats or rooms for residential purposes shall be designed and constructed in such a way as to prevent more reverberation around the common parts than is reasonable.	Requirement E3 only applies to corridors, stairwells, hallways and entrance halls which give access to the flat or room for residential purposes.
Acoustic conditions in schools	
E4. (1) Each room or other space in a school building shall be designed and constructed in such a way that it has the acoustic conditions and the insulation against disturbance by noise appropriate to its intended use. (2) For the purposes of this Part – 'school' has the same meaning as in Section 4 of the Education Act 1996[4]; and 'school building' means any building forming a school or part of a school.	

[4] 1996 c.56. Section 4 was amended by Schedule 22 to the Education Act 1997 (c. 44).

Attention is drawn to the following extracts from the Building Regulations 2010.

Interpretation (Regulation 2) 'room for residential purposes' means a room, or a suite of rooms, which is not a dwelling-house or a flat and which is used by one or more persons to live and sleep and includes a room in a hostel, an hotel, a boarding house, a hall of residence or a residential home, but does not include a room in a hospital, or other similar establishment, used for patient accommodation.

Meaning of material change of use (Regulation 5)

For the purposes of paragraph 8 (1)(e) of Schedule 1 to the Act and for the purposes of these Regulations, there is a material change of use where there is a change in the purposes for which or the circumstances in which a building is used, so that after the change:

a. the building is used as a dwelling, where previously it was not;

b. the building contains a flat, where previously it did not;

c. the building is used as an hotel or boarding house, where previously it was not;

d. the building is used as an institution, where previously it was not;

e. the building is used as a public building, where previously it was not;

f. the building is not a building described in Classes 1 to 6 in Schedule 2, where previously it was;

g. the building, which contains at least one dwelling, contains a greater or lesser number of dwellings than it did previously;

h. the building contains a room for residential purposes, where previously it did not; or

i. the building, which contains at least one room for residential purposes, contains a greater or lesser number of such rooms than it did previously.

j. the building is used as a shop, where previously it was not.

Requirements relating to material change of use (Regulation 6)

1. Where there is a material change of use of the whole of a building, such work, if any, shall be carried out as is necessary to ensure that the building complies with the applicable requirements of the following paragraphs of Schedule 1:

a. in all cases,

B1 (means of warning and escape)

B2 (internal fire spread – linings)

B3 (internal fire spread – structure)

B4(2) (external fire spread – roofs)

B5 (access and facilities for the fire service)

C2(c) (interstitial and surface condensation)

F1 (ventilation)

G1 (cold water supply)

G3(1) to (3) (hot water supply and systems)

G4 (sanitary conveniences and washing facilities)

G5 (bathrooms)

G6 (kitchens and food preparation areas)

H1 (foul water drainage)

H6 (solid waste storage)

J1 to J4 (combustion appliances)

L1 (conservation of fuel and power – dwellings)

P1 (electrical safety);

b. in the case of a material change of use described in Regulations 5(c),(d), (e) or (f), A1 to A3 (structure);

c. in the case of a building exceeding fifteen metres in height, B4(1) (external fire spread – walls);

d. in the case of a material change of use described in Regulation 5(a), (b), (c), (d), (g), (h), (i) or, where the material change of use provides new residential accommodation, (f), C1 (2) (resistance to contaminants);

e. in the case of material change of use described in Regulation 5(a), C2 (resistance to moisture);

f. in the case of a material change of use described in Regulation 5(a), (b), (c), (g), (h) or (i) E1 to E3;

g. in the case of a material change of use described in Regulation 5(e), where the public building consists of or contains a school, E4 (acoustic conditions in schools);

h. in the case of a material change of use described in regulation 5(a) or (b), G2 (water efficiency) and G3(4) (hot water supply and systems: hot water supply to fixed baths);

i. in the case of a material change of use described in regulation 5(c), (d), (e) or (j), M1 (access and use).

2. Where there is a material change of use of part only of a building, such work, if any, shall be carried out as is necessary to ensure that:

a. that part complies in all cases with any applicable requirement referred to in paragraph (1)(a);

b. in a case to which sub-paragraphs (b), (e), (f) or (g) of paragraph (1) apply, that part complies with the requirements referred to in the relevant sub-paragraphs;

c. in the case to which sub-paragraph (c) of paragraph (1) applies, the whole building complies with the requirement referred to in that sub-paragraph; and

d. in a case to which sub-paragraph (i) of paragraph (1) applies –

   i. that part and any sanitary conveniences provided in or in connection with that part comply with the requirements referred to in that sub-paragraph; and

   ii. the building complies with requirement M1(a) of Schedule 1 to the extent that reasonable provision is made to provide either suitable independent access to that part or suitable access through the building to that part.

Sound insulation testing (Regulation 41)

41.

1. Subject to paragraph (4) below, this regulation applies to:

a. building work in relation to which paragraph E1 of Schedule 1 imposes a requirement; and

b. work which is required to be carried out to a building to ensure that it complies with paragraph E1 of Schedule 1 by virtue of Regulation 6(1)(f) or 6(2)(b).

2. Where this Regulation applies, the person carrying out the work shall, for the purpose of ensuring compliance with paragraph E1 of Schedule 1:

a. ensure that appropriate sound insulation testing is carried out in accordance with a procedure approved by the Secretary of State; and

b. give a copy of the results of the testing referred to in sub-paragraph (a) to the local authority.

3. The results of testing referred to in paragraph (2)(a) shall be:

a. recorded in a manner approved by the Secretary of State; and

b. given to the local authority in accordance with paragraph (2)(b) not later than the date on which the notice required by regulation 16(4) is given.

4. Where building work consists of the erection of a dwelling-house or a building containing flats, this regulation does not apply to any part of the building in relation to which the person carrying out the building work notifies the local authority, not later than the date on which he gives notice of commencement of the work under Regulation 16(1), that for the purpose of achieving compliance of the work with paragraph E1 of Schedule 1 he is using one or more design details approved by Robust Details Limited[a], provided that:

a. the notification specifies:

   i. the part or parts of the building in respect of which he is using the design detail;

   ii. the design detail concerned; and

   iii. the unique number issued by Robust Details Limited in respect of the specified use of that design detail; and

b. the building work carried out in respect of the part or parts of the building identified in the notification is in accordance with the design detail specified in the notification.

Attention is drawn to the following extract from the Building (Approved Inspectors etc.) Regulations 2010 (SI 2010/2215)

Sound insulation testing (Regulation 20(1) and (5)

Application of regulations 20, 27, 29, 37, 41, 42, 43 and 44 of the Principal

20.—(1) Regulations 20 (provisions applicable to self-certification schemes), 27 (CO_2 emission rate calculations), 29 (energy performance certificates), 37 (wholesome water consumption calculation), 41 (sound insulation testing), 42 (mechanical ventilation air flow rate testing), 43 (pressure testing) and 44 (commissioning) of the Principal Regulations apply in relation to building work which is the subject of an initial notice as if references to the local authority were references to the approved inspector.

(5) Regulation 41 of the Principal Regulations applies in relation to building work which is the subject of an initial notice as if –

a. for paragraph (3)(b) there were substituted – "(b) given to the approved inspector in accordance with paragraph (2)(b) not later than five days after completion of the work to which the initial notice relates.";

b. for the words in paragraph (4) "not later than the date on which notice of commencement of the work is given under regulation 16(1)" there were substituted the words "prior to the commencement of the building work on site".

[a] A company incorporated under the Companies Acts with the registration number 04980223.

For the purposes of Approved Document E the following definitions apply:

'Adjoining': Adjoining dwelling-houses, adjoining flats, adjoining rooms for residential purposes and adjoining buildings are those in direct physical contact with another dwelling-house, flat, room for residential purposes or building.

'Historic buildings': Historic buildings include:

a. listed buildings

b. buildings situated in conservation areas

c. buildings which are of architectural and historical interest and which are referred to as a material consideration in a local authority's development plan

d. buildings of architectural and historical interest within national parks, areas of outstanding natural beauty, and world heritage sites

e. vernacular buildings of traditional form and construction.

Section 0: Performance

Performance standards

0.1 In the Secretary of State's view the normal way of satisfying Requirement E1 will be to build separating walls, separating floors, and stairs that have a separating function, together with the associated flanking construction, in such a way that they achieve the sound insulation values for dwelling-houses and flats set out in Table 1a, and the values for rooms for residential purposes (see definition in Regulation 2) set out in Table 1b. For walls that separate rooms for residential purposes from adjoining dwelling-houses and flats, the performance standards given in Table 1a should be achieved.

0.2 Regulation 41 of the Building Regulations 2010 and Regulation 20(1) and (5) of the Building (Approved Inspectors, etc.) Regulations 2010 apply to building work to which Requirement E1 applies, and require appropriate sound insulation testing to be carried out. The exception is that, in the case of new-build houses and buildings containing flats, Regulations 41 and 20(1) and (5) do not apply to any relevant part of the building where the design embodies a design detail or details from the set approved and published by Robust Details Ltd; a valid notification is given to the building control body; and the actual work complies with the detail or details specified in the notification. Subject to this exception, which is further explained in *Annex E: Design details approved by Robust Details Ltd*, Regulation 44 applies where building control is being carried out by a local authority, and Regulation 20(1) and (5) applies where it is being carried out by an Approved Inspector. The normal way of satisfying Regulation 41 or 20(1) and (5) will be to implement a programme of sound insulation testing according to the guidance set out in Section 1: Pre-completion testing, of this Approved Document. It is possible for a builder to opt to use design details approved by Robust Details Ltd in some only of the relevant separating structures in a new house or building containing flats, with the other relevant separating structures remaining subject to testing under Regulation 41 or 20(1) and (5). However, it is recommended that expert advice is taken to ensure compatibility of the constructions.

Table 0.1a **Dwelling-houses and flats – performance standards for separating walls, separating floors, and stairs that have a separating function**

	Airborne sound insulation sound insulation $D_{nT,w} + C_{tr}$ dB (Minimum values)	Impact sound insulation $L'_{nT,w}$ dB (Maximum values)
Purpose built dwelling-houses and flats		
Walls	45	-
Floors and stairs	45	62
Dwelling-houses and flats formed by material change of use		
Walls	43	-
Floors and stairs	43	64

Table 0.1b **Rooms for residential purposes – performance standards for separating walls, separating floors, and stairs that have a separating function**

	Airborne sound insulation sound insulation $D_{nT,w} + C_{tr}$ dB (Minimum values)	Impact sound insulation $L'_{nT,w}$ dB (Maximum values)
Purpose built rooms for residential purposes		
Walls	43	-
Floors and stairs	45	62
Rooms for residential purposes formed by material change of use		
Walls	43	-
Floors and stairs	43	64

Table 0.2 Laboratory values for new internal walls and floors within dwelling-houses, flats and rooms for residential purposes, whether purpose built or formed by material change of use

	Airborne sound insulation R_w dB (Minimum values)
Walls	40
Floors	40

0.3 The sound insulation testing should be carried out in accordance with the procedure described in Annex B of this Approved Document, which is the procedure formally approved by the Secretary of State for the purpose of paragraph (2)(a) of Regulation 41 and paragraph (2)(a) of Regulation 20(1) and (5). The results of the testing must be recorded in the manner described in paragraph 1.41 of Section 1 of this Approved Document, which is the manner approved by the Secretary of State for the purposes of paragraph (3)(a) of Regulation 41 and paragraph (3)(a) of Regulation 20(1) and (5). The test results must be given to the building control body in accordance with the time limits set down in Regulation 41 (for cases where building control is being done by the local authority) or Regulation 20(1) and (5) (in cases where it is being done by an Approved Inspector).

0.4 The person carrying out the building work should arrange for sound insulation testing to be carried out by a test body with appropriate third party accreditation. Test bodies conducting testing should preferably have UKAS accreditation (or a European equivalent) for field measurements. The DCLG also regards members of the ANC Registration Scheme as suitably qualified to carry out pre-completion testing.

0.5 Sections 2, 3, 4 and 6 of this Approved Document give examples of constructions which, if built correctly, should achieve the sound insulation values for dwelling-houses and flats set out in Table 1a, and the values for rooms for residential purposes set out in Table 1b. The guidance in these sections is not exhaustive and other designs, materials or products may be used to achieve the required performance.

0.6 Buildings constructed from sub-assemblies that are delivered newly made or selected from stock are no different from any other new building and must comply with all requirements in Schedule 1 of the Building Regulations 2010. In some applications, such as buildings that are constructed to be temporary dwelling-houses, flats, rooms for residential purposes, or school buildings, the provision of reasonable resistance to the passage of sound may vary depending upon the circumstances in the particular case. For example, (a) a building created by dismantling, transporting and re-erecting the sub-assemblies on the same premises would normally be considered to meet the requirements, (b) a building constructed from sub-assemblies obtained from other premises or from stock manufactured before 1 July 2003 would normally be considered to meet the requirements if it satisfies the relevant requirements of Part E that were applicable in 1992 or, for school buildings, the relevant provisions relating to acoustics set out in the 1997 edition of Building Bulletin 87 (ISBN 011271013 1).

0.7 In the case of some historic buildings undergoing a material change of use, it may not be practical to improve the sound insulation to the standards set out in Tables 1a and 1b. The need to conserve the special characteristics of such historic buildings needs to be recognised[1], and in such work, the aim should be to improve sound insulation to the extent that it is practically possible, always provided that the work does not prejudice the character of the historic building, or increase the risk of long-term deterioration to the building fabric or fittings. In arriving at an appropriate balance between historic building conservation and improving sound insulation it would be appropriate to take into account the advice of the local planning authority's conservation officer. In such cases it will be reasonable to improve the sound insulation as much as is practical, and to affix a notice showing the sound insulation value(s) obtained by testing in accordance with Regulation 41 or 20(1) and (5), in a conspicuous place inside the building.

0.8 The performance standards set out in Tables 1a and 1b are appropriate for walls, floors and stairs that separate spaces used for normal domestic purposes. A higher standard of sound insulation may be required between spaces used for normal domestic purposes and communal or non-domestic purposes. In these situations the appropriate level of sound insulation will depend on the noise generated in the communal or non-domestic space. Specialist advice may be needed to establish if a higher standard of sound insulation is required and, if so, to determine the appropriate level.

[1] BS 7913 *The principles of the conservation of historic buildings*, 1998 provides guidance on the principles that should be applied when proposing work on historic buildings.

0.9 In the Secretary of State's view the normal way of satisfying Requirement E2 will be to use constructions for new walls and floors within a dwelling-house, flat or room for residential purposes (including extensions), that provide the laboratory sound insulation values set out in Table 2. Test bodies conducting testing should preferably have UKAS accreditation (or a European equivalent) for laboratory measurements. It is not intended that performance should be verified by testing on site.

0.10 Section 5 gives examples of constructions that should achieve the laboratory values set out in Table 2. The guidance in these sections is not exhaustive and other designs, materials or products may be used to achieve the required performance.

0.11 In the Secretary of State's view the normal way of satisfying Requirement E3 will be to apply the sound absorption measures described in Section 7 of this Approved Document, or other measures of similar effectiveness.

0.12 In the Secretary of State's view the normal way of satisfying Requirement E4 will be to meet the values for sound insulation, reverberation time and indoor ambient noise which are given in Building Bulletin 93 *Acoustic design of schools: performance standards,* published by the Department for Education and available on the internet at www.gov.uk.

0.13 Diagrams 0.1 to 0.3 illustrate the relevant parts of the building that should be protected from airborne and impact sound in order to satisfy Requirements E1 and E2.

Diagram 0.1 **Requirement E1**

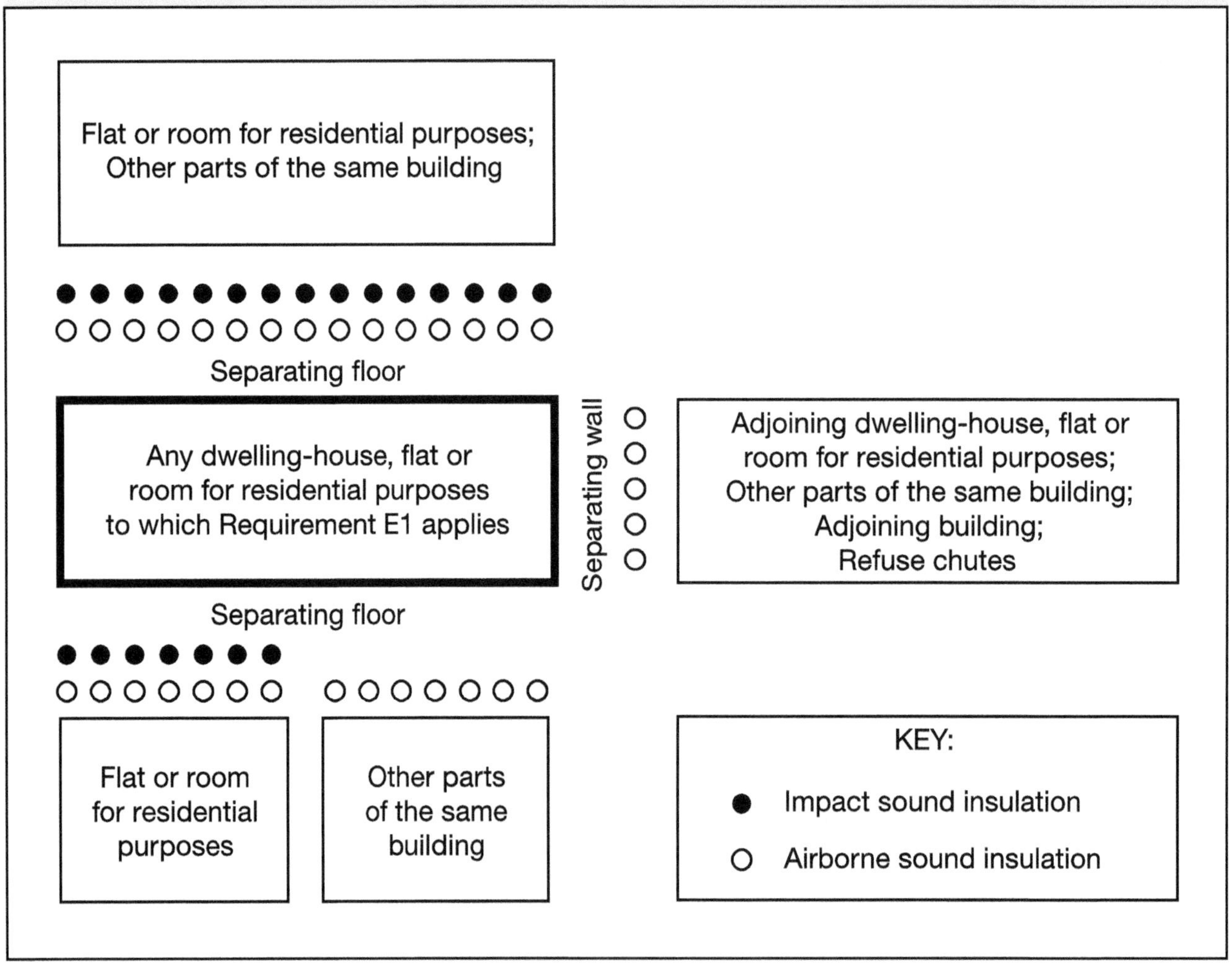

Diagram 0.2 **Requirement E2(a)**

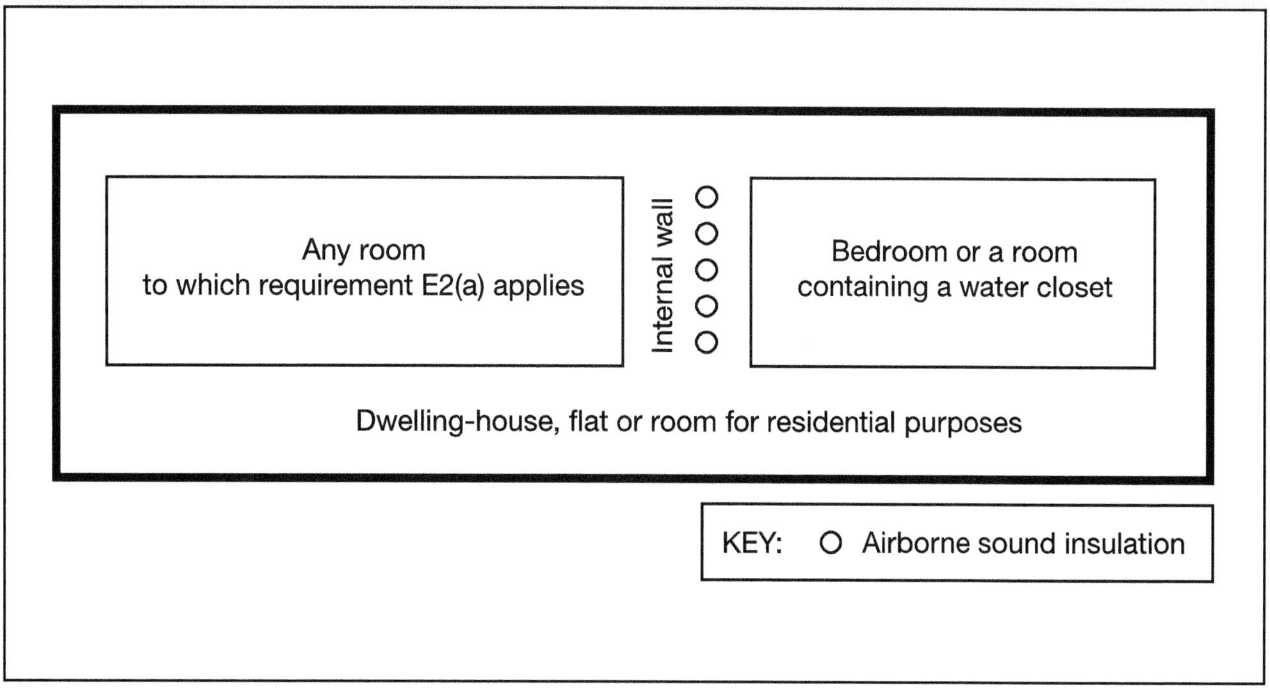

Diagram 0.3 **Requirement E2(b)**

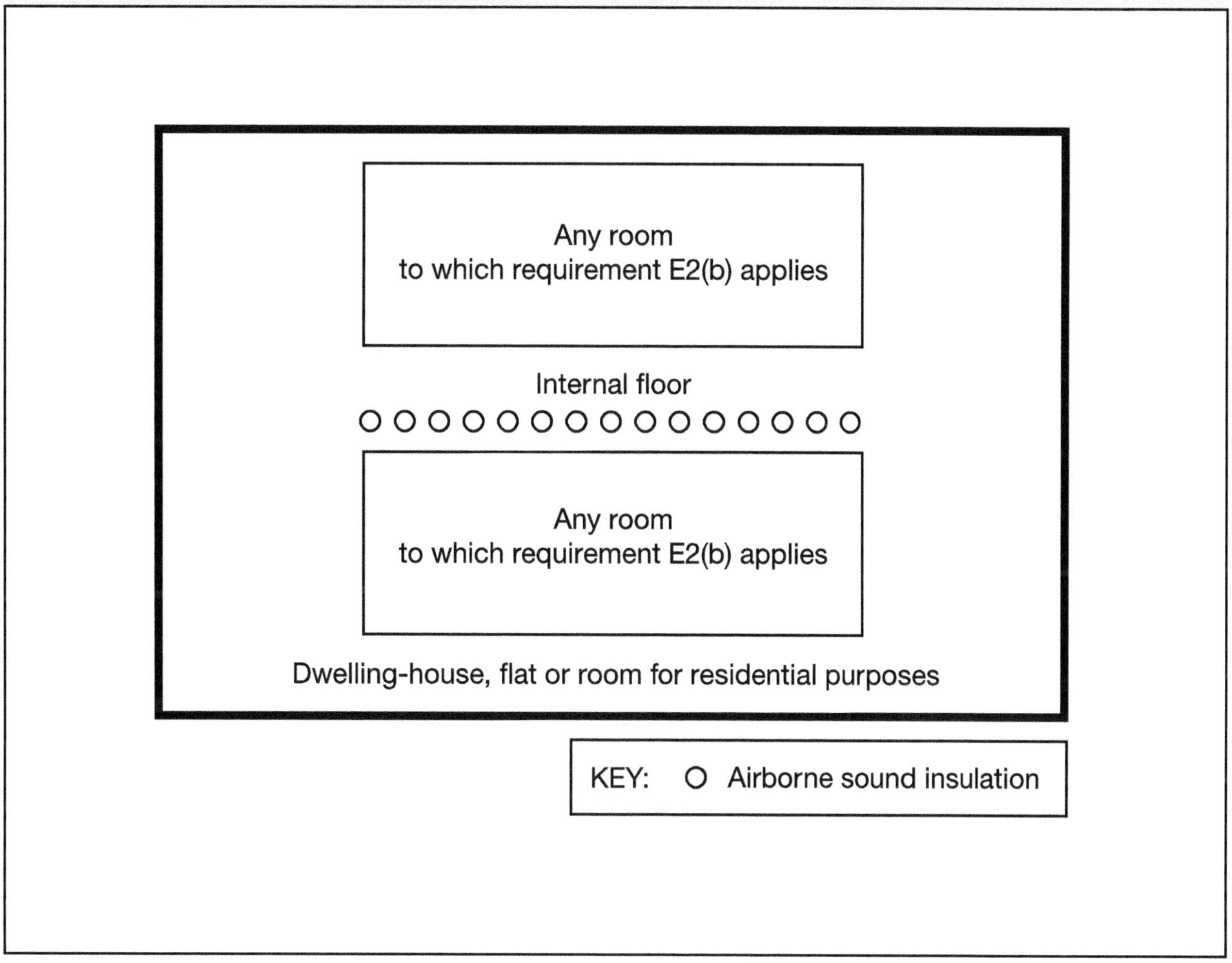

Section 1: Pre-completion testing

Introduction

1.1 This section provides guidance on an appropriate programme of sound insulation testing for a sample of properties, under Regulation 41 of the Building Regulations and Regulation 20(1) and (5) of the Approved Inspectors Regulations.

1.2 Sound insulation testing to demonstrate compliance with Requirement E1 should be carried out on site as part of the construction process, and in this Approved Document it is referred to as pre-completion testing. Under Regulation 41 and Regulation 20(1) and (5), the duty of ensuring that appropriate sound insulation testing is carried out falls on the person carrying out the building work, who is also responsible for the cost of the testing. Therefore, the guidance in this section is addressed in the first place to persons carrying out the work (and to testing bodies employed by them). However, it is also addressed to building control bodies, as the Secretary of State expects building control bodies to determine, for each relevant development, the properties selected for testing.

1.3 Testing should be carried out for:

a. purpose built dwelling-houses and flats;

b. dwelling-houses and flats formed by material change of use;

c. purpose built rooms for residential purposes;

d. rooms for residential purposes formed by material change of use.

1.4 The normal programme of testing is described in paragraphs 1.29 to 1.31.

1.5 The testing procedure formally approved by the Secretary of State is described in Annex B: Procedures for sound insulation testing.

1.6 The performance standards that should be demonstrated by pre-completion testing are set out in Section 0: Performance – Tables 1a and 1b. The sound insulation values in these tables have a built-in allowance for measurement uncertainty, so if any test shows one of these values not to have been achieved by any margin, the test has been failed.

1.7 The person carrying out the building work should ensure that the guidance on construction given in this Approved Document, or in another suitable source, is followed properly to minimise the chances of a failed test. Where additional guidance is required, specialist advice on the building design should be sought at an early stage.

1.8 Testing should not be carried out between living spaces and: corridors, stairwells or hallways.

1.9 Tests should be carried out between rooms or spaces that share a common area of separating wall or separating floor.

1.10 Tests should be carried out once the dwelling-houses, flats or rooms for residential purposes either side of a separating element are essentially complete, except for decoration. Impact sound insulation tests should be carried out without a soft covering (e.g. carpet, foam backed vinyl) on the floor. For exceptions and further information on floor coverings and testing see Annex B: paragraphs B2.13 and B2.14.

Grouping

1.11 The results of tests only apply to the particular constructions tested but are indicative of the performance of others of the same type in the same development. Therefore, in order for meaningful inferences to be made from tests, it is essential that developments are considered as a number of notional groups, with the same construction type within each group.

1.12 Grouping should be carried out according to the following criteria. Dwelling-houses (including bungalows), flats and rooms for residential purposes should be considered as three separate groups. In addition, if significant differences in construction type occur within any of these groups, sub-groups should be established accordingly.

1.13 The following guidance should allow suitable sub-grouping in most circumstances.

Sub-grouping for new buildings

1.14 For dwelling-houses (including bungalows), sub-grouping should be by type of separating wall. For flats, sub-grouping should be by type of separating floor and type of separating wall. Rooms for residential purposes should be grouped using similar principles.

1.15 The construction of flanking elements (e.g. walls, floors, cavities) and their junctions are also important. Where there are significant differences between flanking details, further sub-grouping will be necessary.

1.16 Sub-grouping may not be necessary for dwelling-houses, flats and rooms for residential purposes that have the same separating wall and/or separating floor construction, with the same associated flanking construction(s), and where the room dimensions and layouts are broadly similar.

1.17 Some dwelling-houses, flats or rooms for residential purposes may be considered to have unfavourable features: an example could be flats with large areas of flanking wall without a window at the gable end. It would be inappropriate for these to be included as part of a group and these should form their own sub-group(s).

Sub-grouping for material change of use

1.18 The same principles as for new buildings apply, but in practice significant differences are more likely to occur between separating wall and/or separating floor constructions as well as the associated flanking construction(s) in a development. More sub-groups may therefore be required, and group sizes may be smaller. Building control bodies should exercise judgement when setting up sub-groups.

Sets of tests in dwelling-houses (including bungalows)

1.19 Normally, one set of tests should comprise two individual sound insulation tests (two airborne tests):

- A test of insulation against airborne sound between one pair of rooms (where possible suitable for use as living rooms) on opposite sides of the separating wall.
- A test of insulation against airborne sound between another pair of rooms (where possible suitable for use as bedrooms) on opposite sides of the separating wall.

Sets of tests in flats with separating floors but without separating walls

1.20 Normally, one set of tests should comprise four individual sound insulation tests (two airborne tests, two impact tests):

- Tests of insulation against both airborne and impact sound between one pair of rooms (where possible suitable for use as living rooms) on opposite sides of the separating floor.
- Tests of insulation against both airborne and impact sound between another pair of rooms (where possible suitable for use as bedrooms) on opposite sides of the separating floor.

Sets of tests in flats with a separating floor and a separating wall

1.21 Normally, one set of tests should comprise six individual sound insulation tests (four airborne tests, two impact tests):

- A test of insulation against airborne sound between one pair of rooms (where possible suitable for use as living rooms) on opposite sides of the separating wall.
- A test of insulation against airborne sound between another pair of rooms (where possible suitable for use as bedrooms) on opposite sides of the separating wall.
- Tests of insulation against both airborne and impact sound between one pair of rooms (where possible suitable for use as living rooms) on opposite sides of the separating floor.
- Tests of insulation against both airborne and impact sound between another pair of rooms (where possible suitable for use as bedrooms) on opposite sides of the separating floor.

1.22 To conduct a full set of tests, access to at least three flats will be required.

Types of rooms for testing

1.23 It is preferable that each set of tests contains individual tests in bedrooms and living rooms.

1.24 Where pairs of rooms on either side of the separating element are different (e.g. a bedroom and a study, a living room and a bedroom), at least one of the rooms in one of the pairs should be a bedroom and at least one of the rooms in the other pair should be a living room.

1.25 Where the layout has only one pair of rooms on opposite sides of the entire area of separating wall or floor between two dwelling-houses, flats or rooms for residential purposes then the number of airborne and impact sound insulation tests set out in paragraphs 1.19 to 1.21 may be reduced accordingly.

1.26 The approved procedure described in Annex B includes requirements relating to rooms.

Sets of tests in rooms for residential purposes

1.27 To conduct a set of tests, the sound insulation between the main rooms should be measured according to the principles set out in this section for new buildings and material change of use, but adapting them to suit the circumstances.

Properties sold before fitting out

1.28 Some properties, for example loft apartments, may be sold before being fitted out with internal walls and other fixtures and fittings. Measurements of sound insulation should be made between the available spaces, according to the principles set out in this section. Steps should be taken to ensure that fitting out will not adversely affect the sound insulation. Some guidance on internal wall and floor constructions is given in Section 5. Junction details between these internal walls and floors and separating walls and floors are described in Sections 2 and 3.

Normal programme of testing

1.29 Building control bodies should consult with developers on likely completion times on site, and ask for one set of tests to be carried out between the first dwelling-houses, flats or rooms for residential purposes scheduled for completion and/or sale in each group or sub-group. This applies regardless of the intended size of the group or sub-group. Therefore if a site comprises only one pair of dwelling-houses, flats or rooms for residential purposes, they should be tested.

1.30 As further properties on a development become ready for testing, building control bodies should indicate at what point(s) they wish any further set(s) of tests to be conducted. Assuming no tests are failed, building control bodies should stipulate at least one set of tests for every ten dwelling-houses, flats or rooms for residential purposes in a group or sub-group.

1.31 Testing should be conducted more frequently at the beginning of a series of completions than towards the end, to allow any potential problems to be addressed at an early stage. However, on large developments testing should be carried out over a substantial part of the construction period.

Action following a failed set of tests

1.32 A set of tests is failed if any of its individual tests of airborne or impact sound insulation do not show sound insulation values equal to or better than those set out in Section 0: Performance – Tables 1a and 1b.

1.33 In the event of a failed set of tests, appropriate remedial treatment should be applied to the rooms that failed the test.

1.34 A failed set of tests raises questions over the sound insulation between other rooms sharing the same separating element in the dwelling-houses, flats or rooms for residential purposes in which the tests were conducted. The developer should demonstrate to the building control body's satisfaction that these rooms meet the performance standards. Normally this would be done by (a) additional testing, and/or (b) applying the appropriate remedial treatment to the other rooms and/or (c) demonstrating that the cause of failure does not occur in other rooms.

1.35 A failed set of tests raises questions over properties between which tests have not been carried out. The developer should demonstrate to the building control body's satisfaction that such properties meet the performance standards. Once a dwelling-house, flat or room for residential purposes is occupied, any action affecting it should be a matter for local negotiation.

1.36 After a failed set of tests, the rate of testing should be increased until the building control body is satisfied that the problem has been solved.

Remedial treatment

1.37 Appropriate remedial treatment should be applied following a failed set of tests. It is essential that remedial work is appropriate to the cause of failure. Guidance is available in BRE Information Paper IP 14/02.

1.38 Where the cause of failure is attributed to the construction of the separating and/or associated flanking elements, other rooms that have not been tested may also fail to meet the performance standards. Therefore, remedial treatment may be needed in rooms other than those in which the tests were conducted.

1.39 Where remedial treatment has been applied, the building control body should be satisfied with its efficacy. Normally this will be assessed through additional sound insulation testing.

Material change of use

1.40 As stated in Section 0, in the case of some historic buildings undergoing a material change of use, it may not always be practical to achieve the sound insulation values set out in Section 0: Performance – Tables 1a and 1b. However, in such cases building control bodies should be satisfied that everything reasonable has been done to improve the sound insulation. Tests should be carried out, and the results displayed as indicated in Section 0, paragraph 0.7.

Approved manner of recording pre-completion testing results

1.41 In order to satisfy the requirements of paragraph (3)(a) of Regulation 41 or Regulation 20(1) and (5), the test report of a set of tests (where set of tests has the meaning given in paragraphs 1.19–1.21 and 1.27) must contain at least the following information, in the order below:

1. Address of building.
2. Type(s) of property. Use the definitions in Regulation 2: dwelling-house, flat, room for residential purposes. State if the building is a historic building (see definition in the section on Requirements of this Approved Document).
3. Date(s) of testing.
4. Organisation carrying out testing, including:
   a. name and address;
   b. third party accreditation number (e.g. UKAS or European equivalent);
   c. name(s) of person(s) in charge of test;
   d. name(s) of client(s).

5. A statement (preferably in a table) giving the following information:
   a. the rooms used for each test within the set of tests;
   b. the measured single-number quantity ($D_{nT,w} + C_{tr}$ for airborne sound insulation and $L'_{nT,w}$ for impact sound insulation) for each test within the set of tests;
   c. the sound insulation values that should be achieved according to the values set out in Section 0: Performance – Table 1a or 1b; and
   d. an entry stating 'Pass' or 'Fail' for each test within the set of tests according to the sound insulation values set out in Section 0: Performance – Table 1a or 1b.
6. Brief details of test, including:
   a. equipment;
   b. a statement that the test procedures in Annex B have been followed. If the procedure could not be followed exactly then the exceptions should be described and reasons given;
   c. source and receiver room volumes (including a statement on which rooms were used as source rooms);
   d. results of tests shown in tabular and graphical form for third octave bands according to the relevant part of the BS EN ISO 140 series and BS EN ISO 717 series, including:
      i. single-number quantities and the spectrum adaptation terms;
      ii. D_{nT} and L'_{nT} data from which the single-number quantities are calculated.

Section 2: Separating walls and associated flanking constructions for new buildings

Introduction

2.1 This section gives examples of wall types which, if built correctly, should achieve the performance standards set out in Section 0: Performance – Table 1a.

2.2 The guidance in this section is not exhaustive and other designs, materials or products may be used to achieve the performance standards set out in Section 0: Performance – Table 1a. Advice should be sought from the manufacturer or other appropriate source.

2.3 The walls are grouped into four main types. See Diagram 2.1.

2.4 **Wall type 1:** *Solid masonry*
The resistance to airborne sound depends mainly on the mass per unit area of the wall.

2.5 **Wall type 2:** *Cavity masonry*
The resistance to airborne sound depends on the mass per unit area of the leaves and on the degree of isolation achieved. The isolation is affected by connections (such as wall ties and foundations) between the wall leaves and by the cavity width.

2.6 **Wall type 3:** *Masonry between independent panels*
The resistance to airborne sound depends partly on the type and mass per unit area of the core, and partly on the isolation and mass per unit area of the independent panels.

2.7 **Wall type 4:** *Framed walls with absorbent material*
The resistance to airborne sound depends on the mass per unit area of the leaves, the isolation of the frames, and the absorption in the cavity between the frames.

2.8 Within each wall type the constructions are ranked, as far as possible, with constructions providing higher sound insulation given first.

Junctions between separating walls and other building elements

2.9 In order for the construction to be fully effective, care should be taken to correctly detail the junctions between the separating wall and other elements, such as floors, roofs, external walls and internal walls. Recommendations are also given for the construction of these elements, where it is necessary to control flanking transmission. Notes and diagrams explain the junction details for each of the separating wall types.

2.10 Table 2.1 indicates the inclusion of guidance in this document on the junctions that may occur between each of the four separating wall types and various attached building elements.

Diagram 2.1 **Types of separating wall**

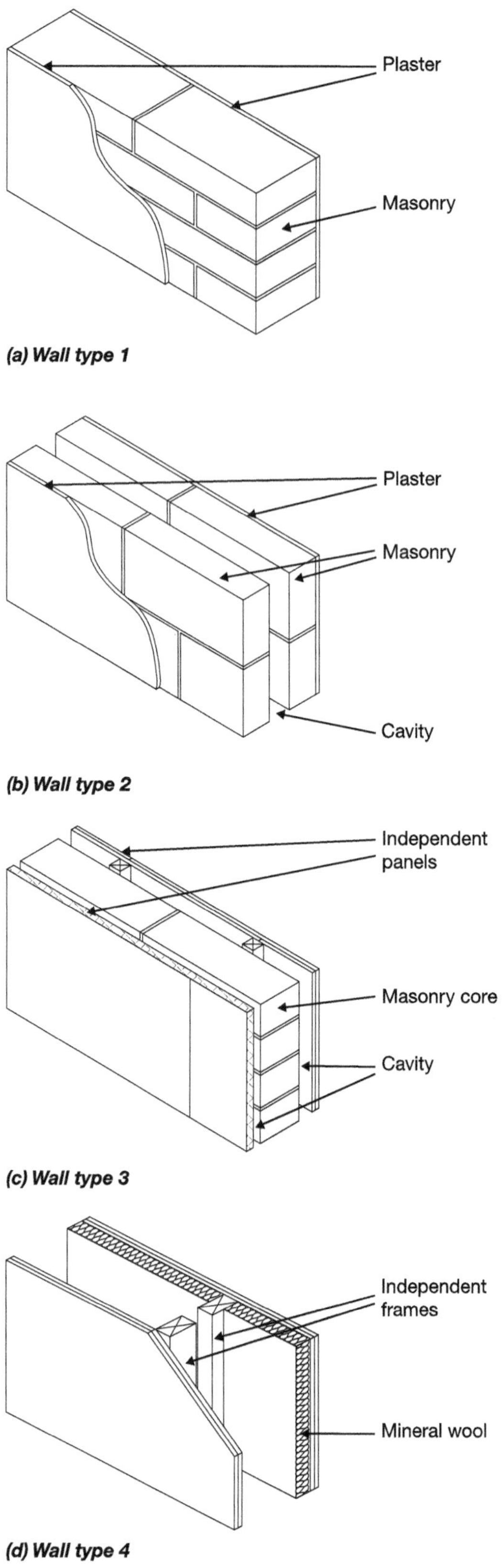

(a) Wall type 1

(b) Wall type 2

(c) Wall type 3

(d) Wall type 4

Table 2.1 **Separating wall junctions reference table**

Building element attached to separating wall	Separating wall type: Type 1	Type 2	Type 3	Type 4
External cavity wall with masonry inner leaf	G	G	G	N
External cavity wall with timber frame inner leaf	G	G	G	G
External solid masonry wall	N	N	N	N
Internal wall – framed	G	G	G	G
Internal wall – masonry	G	G	X	G
Internal floor – timber	G	G	G	G
Internal floor – concrete	G	G	G	N
Ground floor – timber	G	G	G	G
Ground floor – concrete	G	G	G	G
Ceiling and roof space	G	G	G	G
For flats the following may also apply:				
Separating floor type 1 – concrete base with ceiling and soft floor covering	*See Guidance in Section 3, Separating floors and associated flanking constructions for new buildings*			
Separating floor type 2 – concrete base with ceiling and floating floor				
Separating floor type 3 – timber frame base with ceiling and platform floor				

Key: G = guidance available; N = no guidance available (seek specialist advice); X = do not build

Note:

Where any building element functions as a separating element (e.g. a ground floor that is also a separating floor for a basement flat) then the separating element requirements should take precedence.

Mass per unit area of walls

2.11 The mass per unit area of a wall is expressed in kilograms per square metre (kg/m^2). The method for calculating mass per unit area is shown in Annex A.

2.12 The density of the materials used (and on which the mass per unit area of the wall depends) is expressed in kilograms per cubic metre (kg/m^3). When calculating the mass per unit area for bricks and blocks use the density at the appropriate moisture content from Table 3.2, CIBSE Guide A (1999).

2.13 The guidance describes constructions that use blocks without voids. For blocks with voids, seek advice from the manufacturer.

Plasterboard linings on separating and external masonry walls

2.14 The guidance describes some constructions with only wet finishes. For dry finishes, seek advice from the manufacturer.

2.15 Wherever plasterboard is recommended, or the finish is not specified, a drylining laminate of plasterboard with mineral wool may be used. For other drylining laminates, seek advice from the manufacturer.

2.16 Plasterboard linings should be fixed according to manufacturer's instructions.

Cavity widths in separating cavity masonry walls

2.17 Recommended cavity widths are minimum values.

Walls ties in separating and external cavity masonry walls

2.18 Suitable wall ties for use in masonry cavity walls are indicated in the guidance by reference to either tie type A or B.

2.19 Tie type A

Connect the leaves of a cavity masonry wall only where necessary by butterfly ties as described in BS 1243:1978 Metal ties for cavity wall construction, and spaced as required for structural purposes (BS 5628-3:2001 *Code of practice for use of masonry. Materials and components, design and workmanship*, which limits this tie type and spacing to cavity widths of 50mm to 75mm with a minimum masonry leaf thickness of 90mm). Alternatively, use wall ties with an appropriate measured dynamic stiffness for the cavity width. The specification for wall ties of dynamic stiffness, k_{Xmm} in MN/m with a cavity width of Xmm and n ties/m^2 is $n.k_{Xmm}$<4.8MN/m^3.

2.20 Tie type B *(for use only in external masonry cavity walls where tie type A does not satisfy the requirements of Building Regulation Part A – Structure)*

Connect the leaves of a cavity masonry wall only where necessary by double-triangle ties as described in BS 1243:1978 *Metal ties for cavity wall construction*, and spaced as required for structural purposes (BS 5628-3:2001 *Code of practice for use of masonry. Materials and components, design and workmanship*, which limits this tie type and spacing to cavity widths of 50mm to 75mm with a minimum masonry leaf thickness of 90mm). Alternatively, use wall ties with an appropriate measured dynamic stiffness for the cavity width. The specification for wall ties of dynamic stiffness, k_{Xmm} in MN/m with a cavity width of *X*mm and *n* ties/m² is $n.k_{Xmm}$<113MN/m³.

Note: In external cavity masonry walls, tie type B may decrease the airborne sound insulation due to flanking transmission via the external wall leaf compared to tie type A.

2.21 Measurements of the wall tie dynamic stiffness, k_{Xmm}, should be carried out according to BRE Information Paper, IP 3/01.

2.22 The number of ties per square metre, n, is calculated from the horizontal and vertical tie spacing distances, S_x and S_y in metres using $n = 1 / (S_x.S_y)$. Example: for horizontal and vertical tie spacing distances of 0.9m and 0.45m, *n* is 2.5 ties/m².

2.23 If k_{Xmm} is not available for the required cavity width, it is acceptable to use available k_{Xmm} data for *X*mm values less than the required cavity width to calculate $n.k_{Xmm}$.

2.24 All wall ties and spacings specified using the dynamic stiffness parameter should also satisfy the Requirements of Building Regulation Part A – Structure.

Corridor walls and doors

2.25 The separating walls described in this section should be used between corridors and rooms in flats, in order to control flanking transmission and to provide the required sound insulation. However, it is likely that the sound insulation will be reduced by the presence of a door.

2.26 Ensure that any door has good perimeter sealing (including the threshold where practical) and a minimum mass per unit area of 25kg/m² or a minimum sound reduction index of 29 dB R_w (measured according to BS EN ISO 140-3:1995 and rated according to BS EN ISO 717-1:1997). The door should also satisfy the Requirements of Building Regulation Part B – Fire safety.

2.27 Noisy parts of the building should preferably have a lobby, double door or high performance doorset to contain the noise. Where this is not possible, nearby flats should have similar protection. However, there should be a sufficient number of flats that are suitable for disabled access, see Building Regulation Part M – Access and facilities for disabled people.

Refuse chutes

2.28 A wall separating a habitable room or kitchen and a refuse chute should have a mass per unit area (including any finishes) of at least 1320kg/m². A wall separating a non-habitable room from a refuse chute should have a mass per unit area (including any finishes) of at least 220kg/m².

Wall type 1: solid masonry

2.29 The resistance to airborne sound depends mainly on the mass per unit area of the wall.

Constructions

2.30 Three wall type 1 constructions (types 1.1, 1.2, and 1.3) are described in this guidance.

2.31 Details of how junctions should be made to limit flanking transmission are also described in this guidance.

2.32 Points to watch

Do

a. Do fill and seal all masonry joints with mortar.
b. Do lay bricks frog up to achieve the required mass per unit area and avoid air paths.
c. Do use bricks/blocks that extend to the full thickness of the wall.
d. Do ensure that an external cavity wall is stopped with a flexible closer at the junction with a separating wall, unless the cavity is fully filled with mineral wool or expanded polystyrene beads (seek manufacturer's advice for other suitable materials).
e. Do control flanking transmission from walls and floors connected to the separating wall as described in the guidance on junctions.
f. Do stagger the position of sockets on opposite sides of the separating wall.
g. Do ensure that flue blocks will not adversely affect the sound insulation and that a suitable finish is used over the flue blocks (see BS 1289-1:1986 and seek manufacturer's advice).

Do not

a. Do not try and convert a cavity separating wall to a type 1 (solid masonry) separating wall by inserting mortar or concrete into the cavity between the two leaves.
b. Do not use deep sockets and chases in the separating wall, and do not place sockets back to back.
c. Do not create a junction between a solid wall type 1 and a cavity wall type 2 in which the cavity wall is bridged by the solid wall.

2.33 Wall type 1.1 *Dense aggregate concrete block, plaster on both room faces (see Diagram 2.2)*

- minimum mass per unit area including plaster 415kg/m²;
- 13mm plaster on both room faces;
- use blocks that are laid flat to the full thickness of the wall.

Example of wall type 1.1

The required mass per unit area would be achieved by using

- 215mm block laid flat
- block density 1840kg/m³
- 110mm coursing
- 13mm lightweight plaster (minimum mass per unit area 10kg/m²) on both room faces

This is an example only. See Annex A for a simplified method of calculating mass per unit area. Alternatively use manufacturer's actual figures where these are available.

Diagram 2.2 **Wall type 1.1**

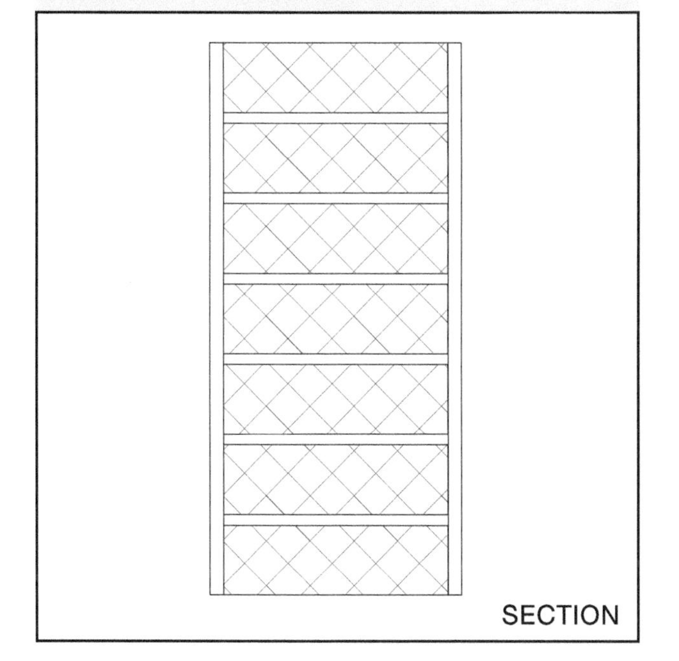

2.34 Wall type 1.2 *Dense aggregate concrete cast in-situ, plaster on both room faces (see Diagram 2.3)*

- minimum mass per unit area including plaster 415kg/m²;
- plaster on both room faces.

Example of wall type 1.2

The required mass per unit area would be achieved by using

- 190mm concrete
- concrete density 2200kg/m³
- 13mm lightweight plaster (minimum mass per unit area 10kg/m²) on both room faces

This is an example only. See Annex A for a simplified method of calculating mass per unit area. Alternatively use manufacturer's actual figures where these are available.

Diagram 2.3 **Wall type 1.2**

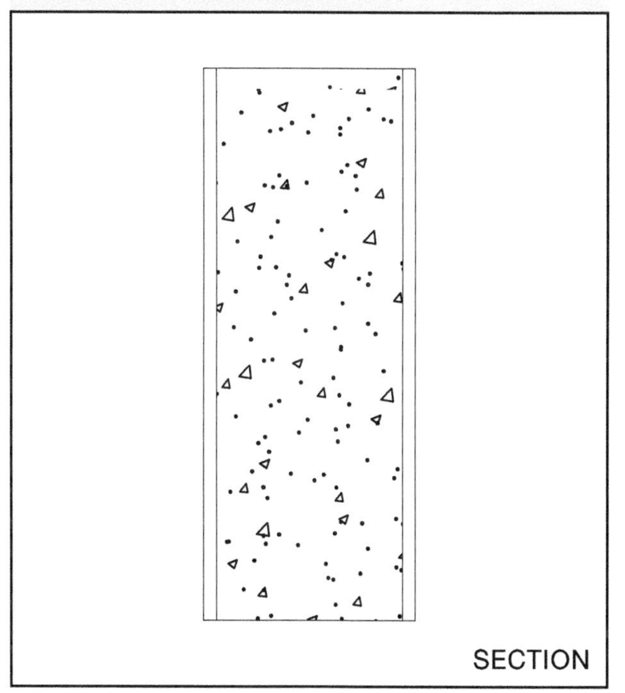

2.35 Wall type 1.3 *Brick, plaster on both room faces (see Diagram 2.4)*

- minimum mass per unit area including plaster 375kg/m²;
- 13mm plaster on both room faces;
- bricks to be laid frog up, coursed with headers.

Example of wall type 1.3

The required mass per unit area would be achieved by using

- 215mm brick
- brick density 1610kg/m³
- 75mm coursing
- 13mm lightweight plaster (minimum mass per unit area 10kg/m²) on both room faces

This is an example only. See Annex A for a simplified method of calculating mass per unit area. Alternatively use manufacturer's actual figures where these are available.

Diagram 2.4 **Wall type 1.3**

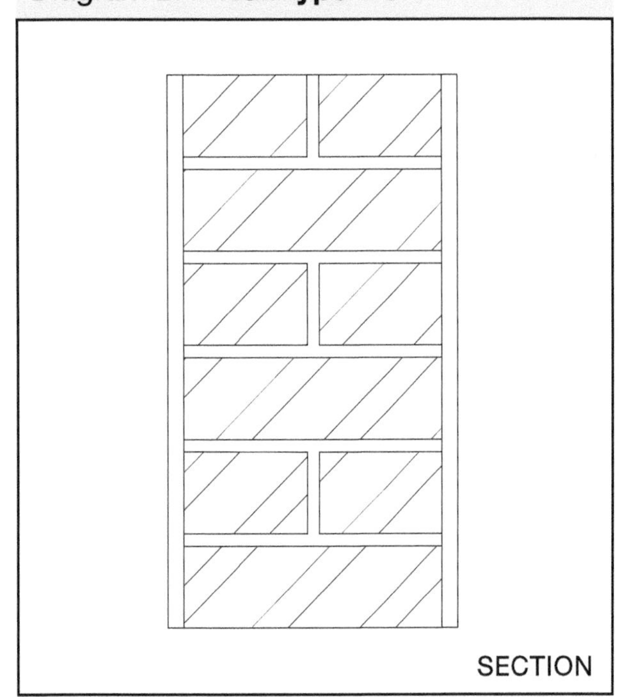

Junction requirements for wall type 1

Junctions with an external cavity wall with masonry inner leaf

2.36 Where the external wall is a cavity wall:

a. the outer leaf of the wall may be of any construction; and

b. the cavity should be stopped with a flexible closer (see Diagram 2.5) unless the cavity is fully filled with mineral wool or expanded polystyrene beads (seek manufacturer's advice for other suitable materials).

2.37 The separating wall should be joined to the inner leaf of the external cavity wall by one of the following methods:

a. Bonded. The separating wall should be bonded to the external wall in such a way that the separating wall contributes at least 50% of the bond at the junction. See Diagram 2.6.

b. Tied. The external wall should abut the separating wall and be tied to it. See Diagram 2.7. Also, see Building Regulation Part A – Structure.

2.38 The masonry inner leaf should have a mass per unit area of at least 120kg/m² excluding finish. However, there is no minimum mass requirement where there are openings in the external wall (see Diagram 2.8) that are:

a. not less than 1 metre high; and

b. on both sides of the separating wall at every storey; and

c. not more than 700mm from the face of the separating wall on both sides.

2.39 Where there is also a separating floor then the requirement for a minimum mass per unit area of 120kg/m² excluding finish should always apply, irrespective of the presence or absence of openings.

Diagram 2.5 **Wall type 1 – external cavity wall with masonry inner leaf**

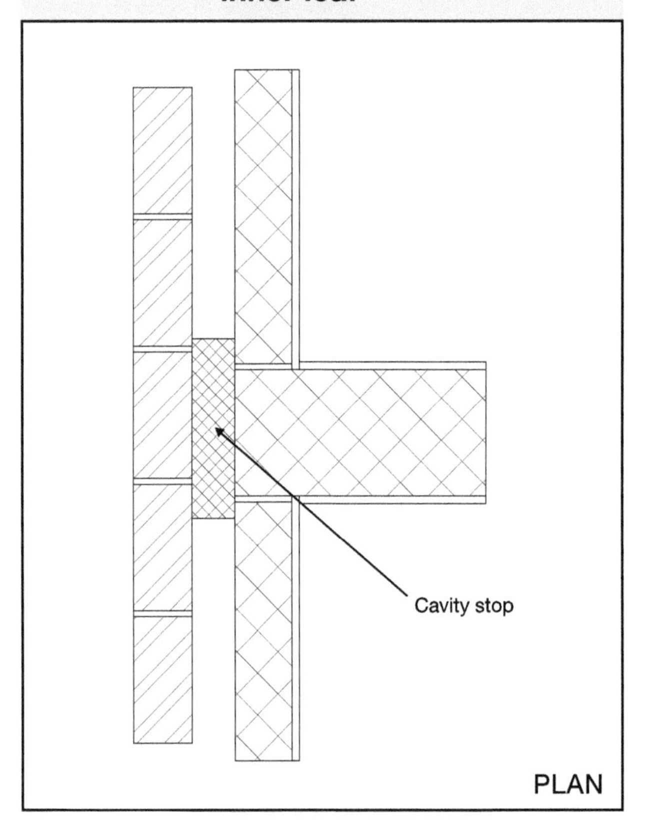

Diagram 2.6 **Wall type 1 – bonded junction – masonry inner leaf of external cavity wall with solid separating wall**

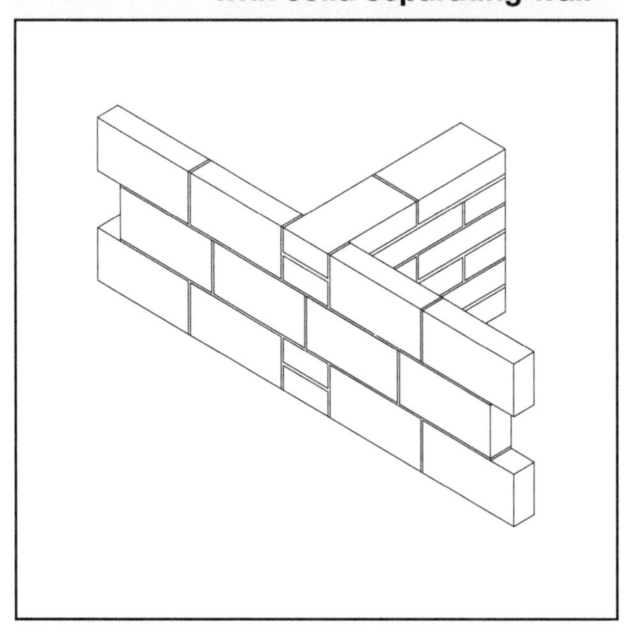

Diagram 2.7 **Wall type 1 – tied junction – external cavity wall with internal masonry wall**

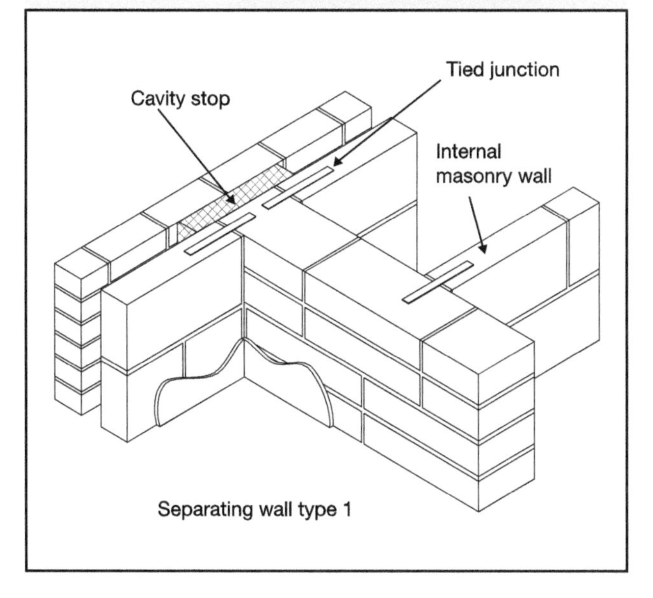

Diagram 2.8 **Wall type 1 – position of openings in masonry inner leaf of external cavity wall**

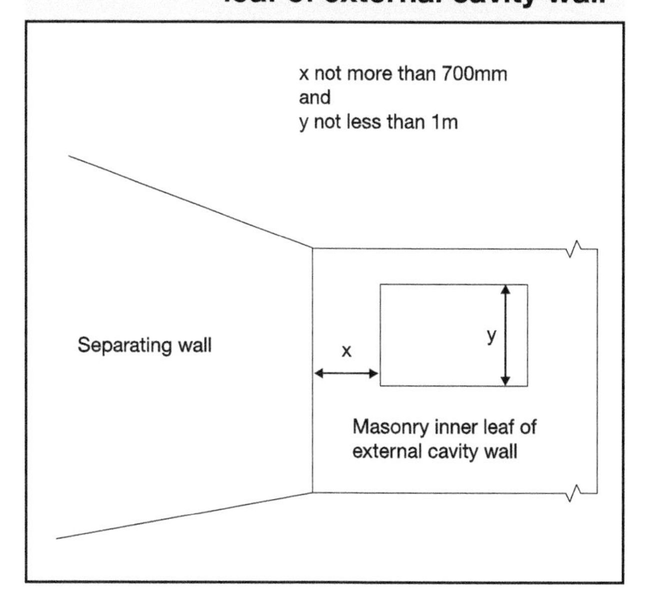

Junctions with an external cavity wall with timber frame inner leaf

2.40 Where the external wall is a cavity wall:

a. the outer leaf of the wall may be of any construction; and

b. the cavity should be stopped with a flexible closer. See Diagram 2.9.

2.41 Where the inner leaf of an external cavity wall is of framed construction, the framed inner leaf should:

a. abut the separating wall; and

b. be tied to it with ties at no more than 300mm centres vertically.

The wall finish of the framed inner leaf of the external wall should be:

a. one layer of plasterboard; or

b. two layers of plasterboard where there is a separating floor;

c. each sheet of plasterboard to be of minimum mass per unit area 10kg/m^2; and

d. all joints should be sealed with tape or caulked with sealant.

Diagram 2.9 **Wall type 1 – external cavity wall with timber frame inner leaf**

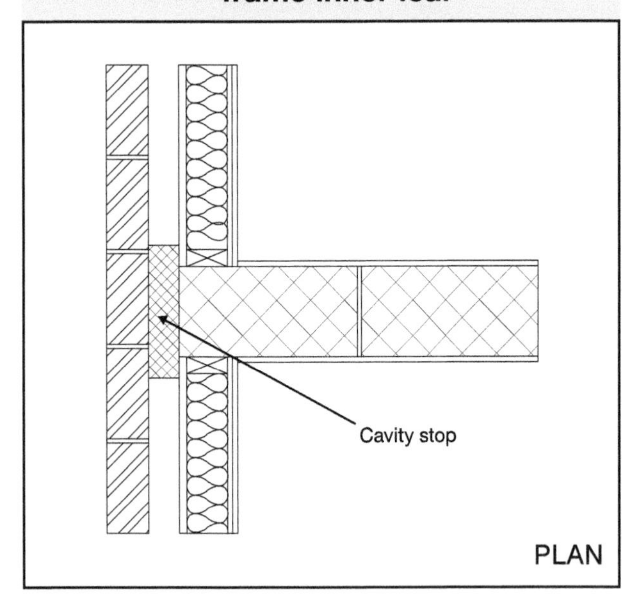

Diagram 2.10 **Wall type 1 – internal timber floor**

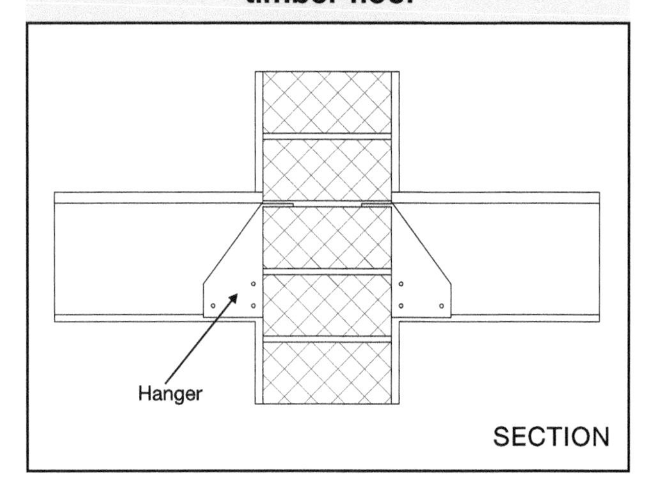

Diagram 2.11 **Wall type 1 – internal concrete floor**

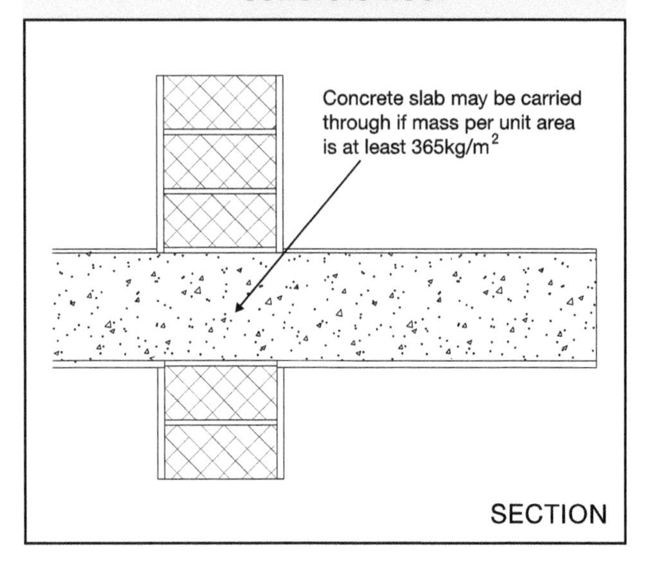

Junctions with an external solid masonry wall

2.42 No guidance available (seek specialist advice).

Junctions with internal framed walls

2.43 There are no restrictions on internal framed walls meeting a type 1 separating wall.

Junctions with internal masonry walls

2.44 Internal masonry walls that abut a type 1 separating wall should have a mass per unit area of at least 120kg/m^2 excluding finish.

Junctions with internal timber floors

2.45 If the floor joists are to be supported on a type 1 separating wall then they should be supported on hangers and should not be built in. See Diagram 2.10.

Junctions with internal concrete floors

2.46 An internal concrete floor slab may only be carried through a type 1 separating wall if the floor base has a mass per unit area of at least 365kg/m^2. See Diagram 2.11.

2.47 Internal hollow-core concrete plank floors and concrete beams with infilling block floors should not be continuous through a type 1 separating wall.

2.48 For internal floors of concrete beams with infilling blocks, avoid beams built in to the separating wall unless the blocks in the floor fill the space between the beams where they penetrate the wall.

Junctions with timber ground floors

2.49 If the floor joists are to be supported on a type 1 separating wall then they should be supported on hangers and should not be built in.

2.50 See Building Regulation Part C – Site preparation and resistance to moisture, and Building Regulation Part L – Conservation of fuel and power.

Junctions with concrete ground floors

2.51 The ground floor may be a solid slab, laid on the ground, or a suspended concrete floor. A concrete slab floor on the ground may be continuous under a type 1 separating wall. See Diagram 2.12.

2.52 A suspended concrete floor may only pass under a type 1 separating wall if the floor has a mass of at least 365kg/m^2.

2.53 Hollow core concrete plank and concrete beams with infilling block floors should not be continuous under a type 1 separating wall.

Diagram 2.12 **Wall type 1 – concrete ground floor**

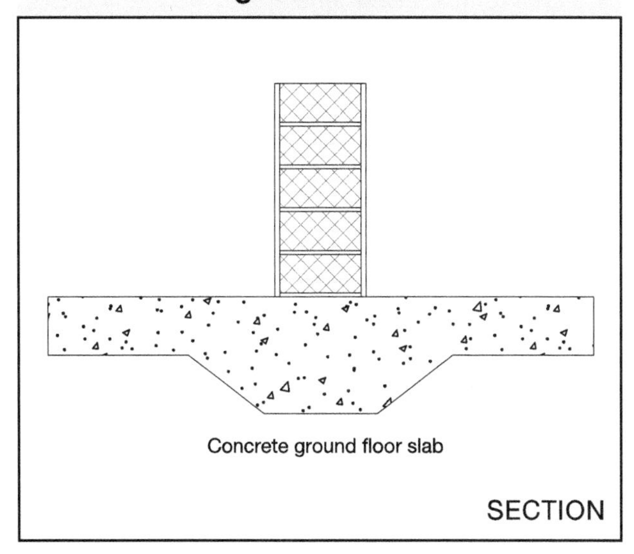

Diagram 2.13 **Wall type 1 – ceiling and roof junction**

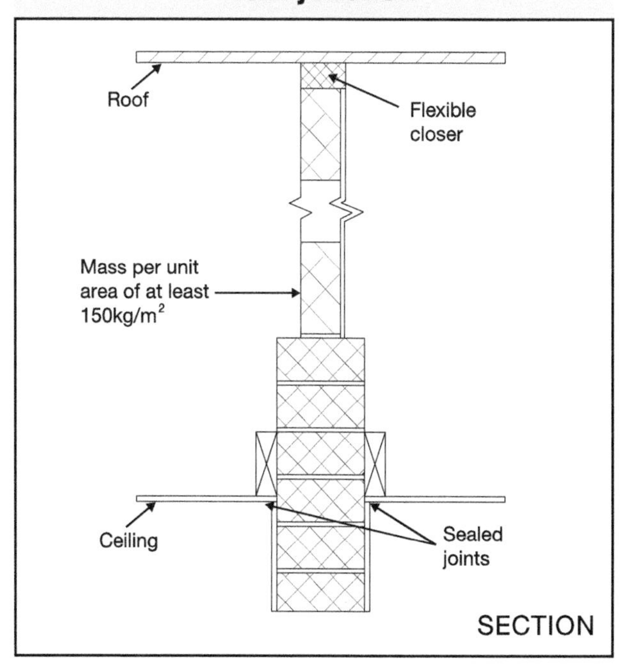

2.54 See Building Regulation Part C – Site preparation and resistance to moisture, and Building Regulation Part L – Conservation of fuel and power.

Junctions with ceiling and roof

2.55 Where a type 1 separating wall is used it should be continuous to the underside of the roof.

2.56 The junction between the separating wall and the roof should be filled with a flexible closer which is also suitable as a fire stop. See Diagram 2.13.

2.57 Where the roof or loft space is not a habitable room and there is a ceiling with a minimum mass per unit area of 10kg/m² with sealed joints, then the mass per unit area of the separating wall above the ceiling may be reduced to 150kg/m². See Diagram 2.13.

2.58 If lightweight aggregate blocks of density less than 1200kg/m³ are used above ceiling level, then one side should be sealed with cement paint or plaster skim.

2.59 Where there is an external cavity wall, the cavity should be closed at eaves level with a suitable flexible material (e.g. mineral wool). See Diagram 2.14.

Note: A rigid connection between the inner and external wall leaves should be avoided. If a rigid material is used, then it should only be rigidly bonded to one leaf. See BRE BR 262, Thermal Insulation: avoiding risks, Section 2.3.

Junctions with separating floors

2.60 There are important details in Section 3 concerning junctions between wall type 1 and separating floors.

Diagram 2.14 **External cavity wall at eaves level**

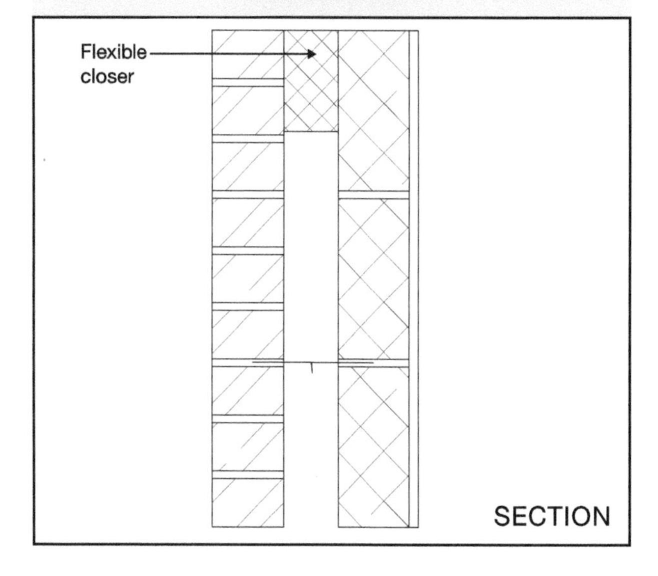

Wall type 2: cavity masonry

2.61 The resistance to airborne sound depends on the mass per unit area of the leaves and on the degree of isolation achieved. The isolation is affected by connections (such as wall ties and foundations) between the wall leaves and by the cavity width.

Constructions

2.62 Four wall type 2 constructions (types 2.1, 2.2, 2.3 and 2.4) are described in this guidance.

2.63 Two of these wall constructions (types 2.3 and 2.4) are only suitable when a step in elevation and/or a stagger in plan is incorporated at the separating wall.

2.64 Details of how junctions should be made to limit flanking transmission are also described in this guidance.

2.65 Points to watch:

Do

a. Do fill and seal all masonry joints with mortar.

b. Do keep the cavity leaves separate below ground floor level.

c. Do ensure that any external cavity wall is stopped with a flexible closer at the junction with the separating wall, unless the cavity is fully filled with mineral wool or expanded polystyrene beads (seek manufacturer's advice for other suitable materials).

d. Do control flanking transmission from walls and floors connected to the separating wall as described in the guidance on junctions.

e. Do stagger the position of sockets on opposite sides of the separating wall.

f. Do ensure that flue blocks will not adversely affect the sound insulation and that a suitable finish is used over the flue blocks (see BS 1289-1:1986 and seek manufacturer's advice).

Do not

a. Do not try and convert a cavity separating wall to a type 1 (solid masonry) separating wall by inserting mortar or concrete into the cavity between the two leaves.

b. Do not change to a solid wall construction in the roof space as a rigid connection between the leaves will reduce wall performance.

c. Do not build cavity walls off a continuous solid concrete slab floor.

d. Do not use deep sockets and chases in the separating wall, do not place them back to back.

Wall ties in separating cavity masonry walls

2.66 The wall ties used to connect the leaves of a cavity masonry wall should be tie type A.

Cavity widths in separating cavity masonry walls

2.67 Recommended cavity widths are minimum values.

Blocks with voids

2.68 The guidance describes constructions that use blocks without voids. For blocks with voids, seek advice from the manufacturer.

2.69 Wall type 2.1 *Two leaves of dense aggregate concrete block with 50mm cavity, plaster on both room faces (see Diagram 2.15)*

- minimum mass per unit area including plaster 415kg/m^2;
- minimum cavity width of 50mm;
- 13mm plaster on both room faces.

Example of wall type 2.1

The required mass per unit area would be achieved by using

- 100mm block leaves
- block density 1990kg/m^3
- 225mm coursing
- 13mm lightweight plaster (minimum mass per unit area 10kg/m^2) on both room faces

This is an example only. See Annex A for a simplified method of calculating mass per unit area. Alternatively use manufacturer's actual figures where these are available.

Diagram 2.15 **Wall type 2.1**

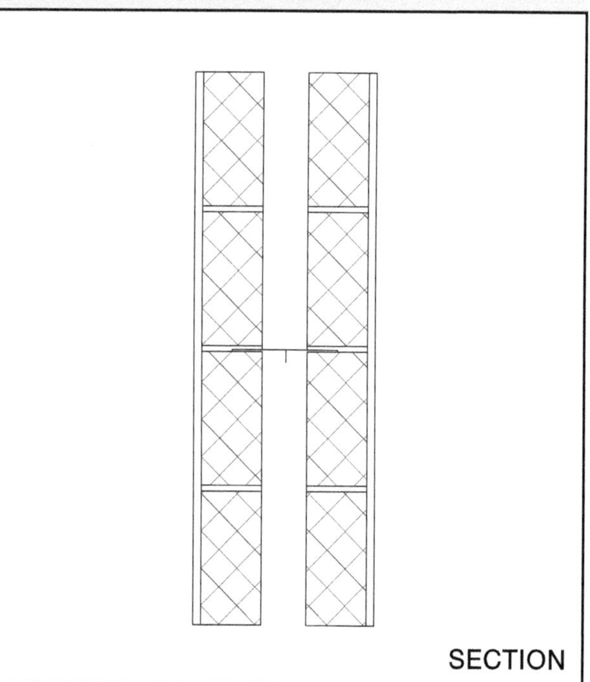

2.70 Wall type 2.2 *Two leaves of lightweight aggregate block with 75mm cavity, plaster on both room faces (see Diagram 2.16)*

- minimum mass per unit area including plaster 300kg/m²;
- minimum cavity width of 75mm;
- 13mm plaster on both room faces.

Example of wall type 2.2

The required mass per unit area would be achieved by using

- 100mm block leaves
- block density 1375kg/m³
- 225mm coursing
- 13mm lightweight plaster (minimum mass per unit area 10kg/m²) on both room faces

This is an example only. See Annex A for a simplified method of calculating mass per unit area. Alternatively use manufacturer's actual figures where these are available.

Diagram 2.16 **Wall type 2.2**

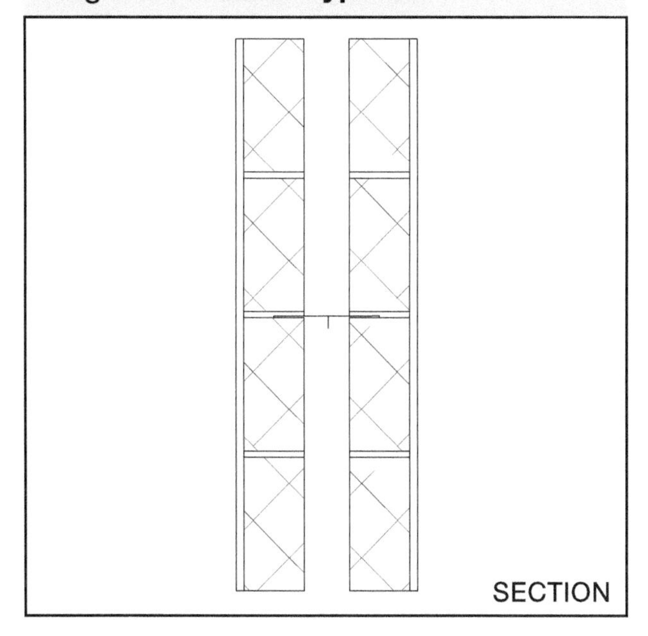

Additional construction: wall type 2.3 should only be used where there is a step and/or stagger of at least 300mm.

2.71 Wall type 2.3 *Two leaves of lightweight aggregate block with 75mm cavity and step/stagger, plasterboard on both room faces (see Diagram 2.17)*

- minimum mass per unit area including plasterboard 290kg/m²;
- lightweight aggregate blocks should have a density in the range 1350 to 1600kg/m³;
- minimum cavity width of 75mm;
- plasterboard, each sheet of minimum mass per unit area 10kg/m², on both room faces.

Note: The composition of the lightweight aggregate blocks contributes to the performance of this construction with a plasterboard finish. Using denser blocks may not give an equivalent performance.

Example of wall type 2.3

The required mass per unit area would be achieved by using

- 100mm block leaves
- block density 1375kg/m³
- 225mm coursing
- plasterboard, each sheet of minimum mass per unit area 10kg/m², on both room faces

This is an example only. See Annex A for a simplified method of calculating mass per unit area. Alternatively use manufacturer's actual figures where these are available.

Note: Increasing the size of the step or stagger in the separating wall tends to increase the airborne sound insulation.

Diagram 2.17 **Wall type 2.3**

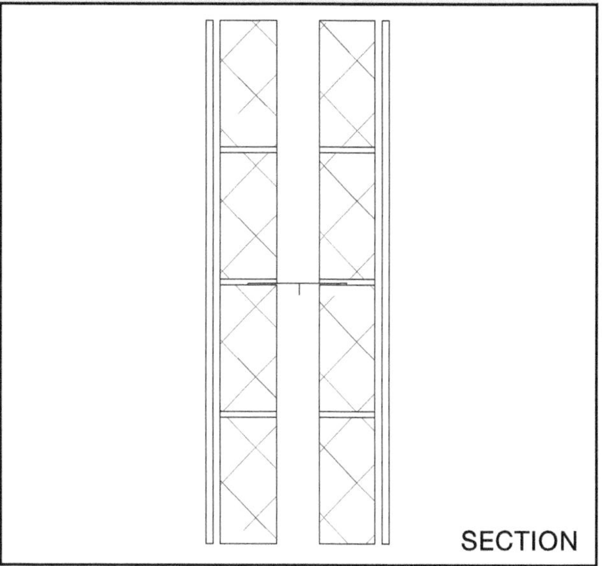

Additional construction: Wall type 2.4 should only be used in constructions without separating floors and where there is a step and/or stagger of at least 300mm.

2.72 Wall type 2.4 Two leaves of aircrete block with 75mm cavity and step/stagger, plasterboard or plaster on both room faces (see Diagram 2.18)

- minimum mass per unit area including finish 150kg/m²;
- minimum cavity width of 75mm;
- plasterboard, each sheet of minimum mass per unit area 10kg/m², on both room faces; or
- 13mm plaster on both room faces.

Example of wall type 2.4

The required mass per unit area would be achieved by using

- 100mm aircrete block leaves
- block density 650kg/m³
- 225mm coursing
- plasterboard, each sheet of minimum mass per unit area 10kg/m², on both room faces

This is an example only. See Annex A for a simplified method of calculating mass per unit area. Alternatively use manufacturer's actual figures where these are available.

Note: Increasing the size of the step or stagger in the separating wall tends to increase the airborne sound insulation.

Diagram 2.18 **Wall type 2.4**

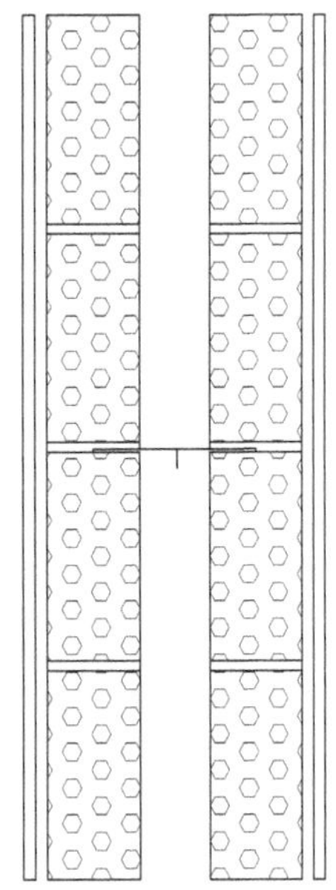

SECTION

Junction requirements for wall type 2

Junctions with an external cavity wall with masonry inner leaf

2.73 Where the external wall is a cavity wall:

a. the outer leaf of the wall may be of any construction; and

b. the cavity should be stopped with a flexible closer (for wall types 2.1 and 2.2 see Diagram 2.19, for wall types 2.3 and 2.4 see Diagram 2.20) unless the cavity is fully filled with mineral wool or expanded polystyrene beads (seek manufacturer's advice for other suitable materials).

2.74 The separating wall should be joined to the inner leaf of the external cavity wall by one of the following methods:

a. Bonded. The separating wall should be bonded to the external wall in such a way that the separating wall contributes at least 50% of the bond at the junction.

b. Tied. The external wall should abut the separating wall and be tied to it. See Diagram 2.21. Also, see Building Regulation Part A – Structure.

2.75 The masonry inner leaf should have a mass per unit area of at least 120kg/m² excluding finish. However, there is no minimum mass requirement where separating wall type 2.1, 2.3 or 2.4 is used.

2.76 Where there is also a separating floor then the requirement for a minimum mass per unit area of 120kg/m² excluding finish should always apply, even when wall type 2.1, 2.3 or 2.4 is used.

Diagram 2.19 **Wall types 2.1 and 2.2 – external cavity wall with masonry inner leaf**

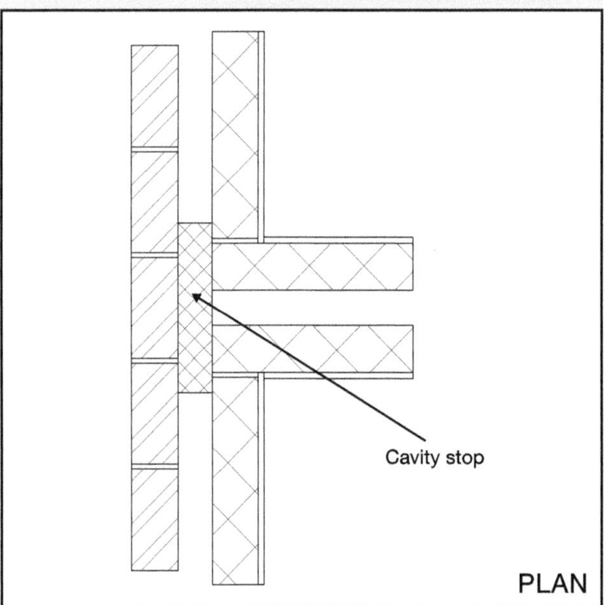

Diagram 2.20 **Wall types 2.3 and 2.4 – external cavity wall with masonry inner leaf – stagger**

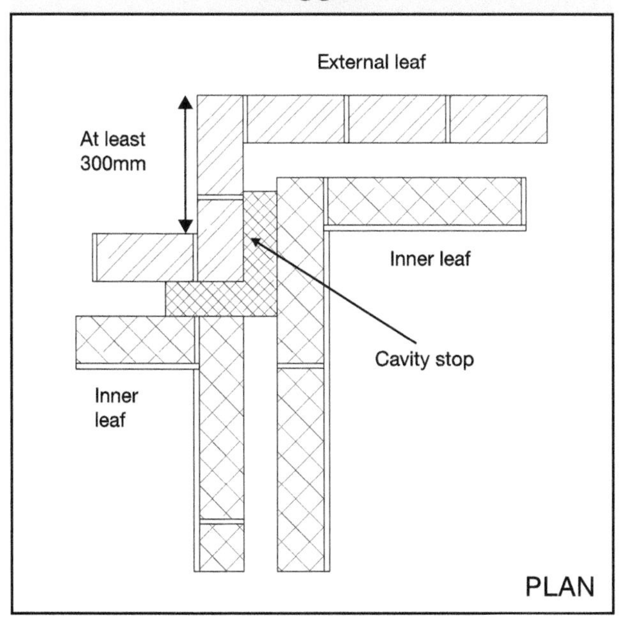

Diagram 2.21 **Wall type 2 – tied junction – external cavity wall with internal masonry wall**

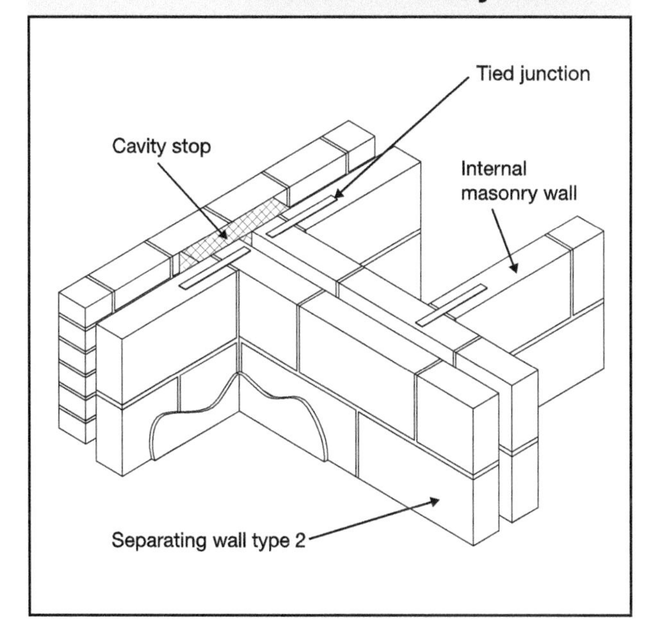

Junctions with an external cavity wall with timber frame inner leaf

2.77 Where the external wall is a cavity wall:

a. the outer leaf of the wall may be of any construction; and

b. the cavity should be stopped with a flexible closer. See Diagram 2.22.

Diagram 2.22 **Wall type 2 – external cavity wall with timber frame inner leaf**

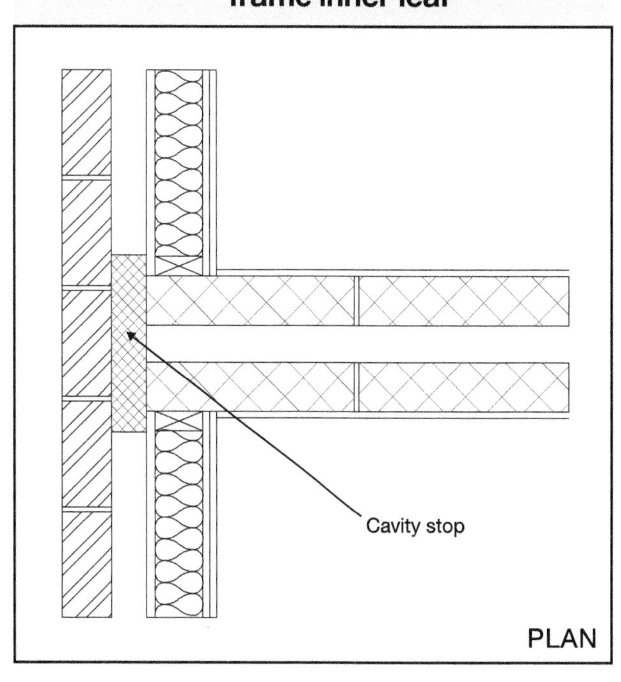

2.78 Where the inner leaf of an external cavity wall is of framed construction, the framed inner leaf should:

a. abut the separating wall; and

b. be tied to it with ties at no more than 300mm centres vertically.

The wall finish of the inner leaf of the external wall should be:

a. one layer of plasterboard; or

b. two layers of plasterboard where there is a separating floor;

c. each sheet of plasterboard to be of minimum mass per unit area 10kg/m²; and

d. all joints should be sealed with tape or caulked with sealant.

Junctions with an external solid masonry wall

2.79 No guidance available (seek specialist advice).

Junctions with internal framed walls

2.80 There are no restrictions on internal framed walls meeting a type 2 separating wall.

Junctions with internal masonry walls

2.81 Internal masonry walls that abut a type 2 separating wall should have a mass per unit area of at least 120kg/m² excluding finish.

2.82 Where there is a separating floor, internal masonry walls should have a mass per unit area of at least 120kg/m² excluding finish.

2.83 When there is no separating floor with separating wall type 2.3 or 2.4 there is no minimum mass per unit area for internal masonry walls.

Junctions with internal timber floors

2.84 If the floor joists are to be supported on the separating wall then they should be supported on hangers and should not be built in. See Diagram 2.23.

Diagram 2.23 **Wall type 2 – internal timber floor**

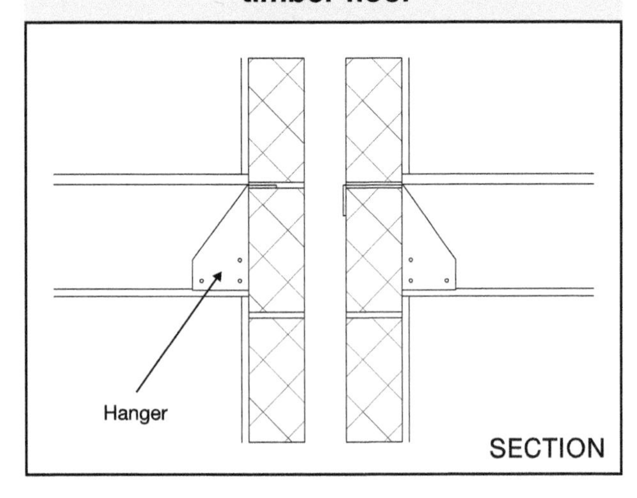

Junctions with internal concrete floors

2.85 Internal concrete floors should generally be built into a type 2 separating wall and carried through to the cavity face of the leaf. The cavity should not be bridged. See Diagram 2.24.

Junctions with timber ground floors

2.86 If the floor joists are to be supported on the separating wall then they should be supported on hangers and should not be built in.

2.87 See Building Regulation Part C – Site preparation and resistance to moisture, and Building Regulation Part L – Conservation of fuel and power.

Junctions with concrete ground floors

2.88 The ground floor may be a solid slab, laid on the ground, or a suspended concrete floor. A concrete slab floor on the ground should not be continuous under a type 2 separating wall. See Diagram 2.24.

2.89 A suspended concrete floor should not be continuous under a type 2 separating wall, and should be carried through to the cavity face of the leaf. The cavity should not be bridged. See Diagram 2.24.

2.90 See Building Regulation Part C – Site preparation and resistance to moisture, and Building Regulation Part L – Conservation of fuel and power.

Diagram 2.24 **Wall type 2 – internal concrete floor and concrete ground floor**

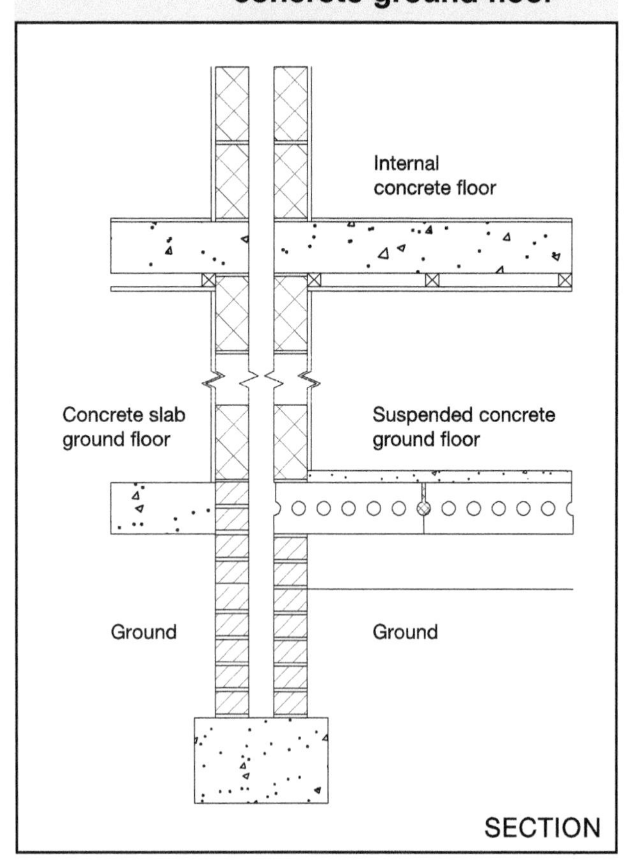

Diagram 2.25 **Wall type 2 – ceiling and roof junction**

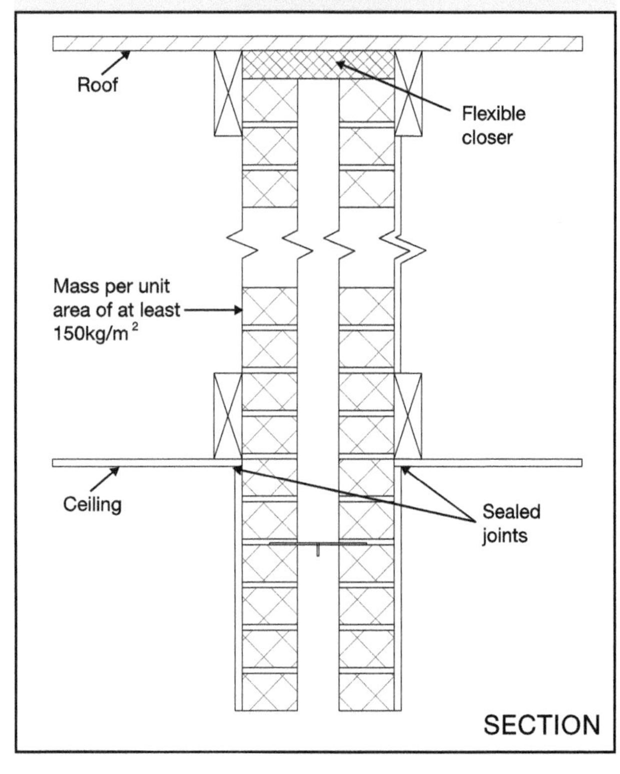

Junctions with ceiling and roof space

2.91 Where a type 2 separating wall is used it should be continuous to the underside of the roof.

2.92 The junction between the separating wall and the roof should be filled with a flexible closer which is also suitable as a fire stop. See Diagram 2.25.

2.93 Where the roof or loft space is not a habitable room and there is a ceiling with a minimum mass per unit area of 10kg/m² with sealed joints, then the mass per unit area of the separating wall above the ceiling may be reduced to 150kg/m² , but it should still be a cavity wall. See Diagram 2.25.

2.94 If lightweight aggregate blocks of density less than 1200kg/m³ are used above ceiling level, then one side should be sealed with cement paint or plaster skim.

2.95 Where there is an external cavity wall, the cavity should be closed at eaves level with a suitable flexible material (e.g. mineral wool). See Diagram 2.26.

Note: A rigid connection between the inner and external wall leaves should be avoided. If a rigid material is used, then it should only be rigidly bonded to one leaf.

Diagram 2.26 **External cavity wall at eaves level**

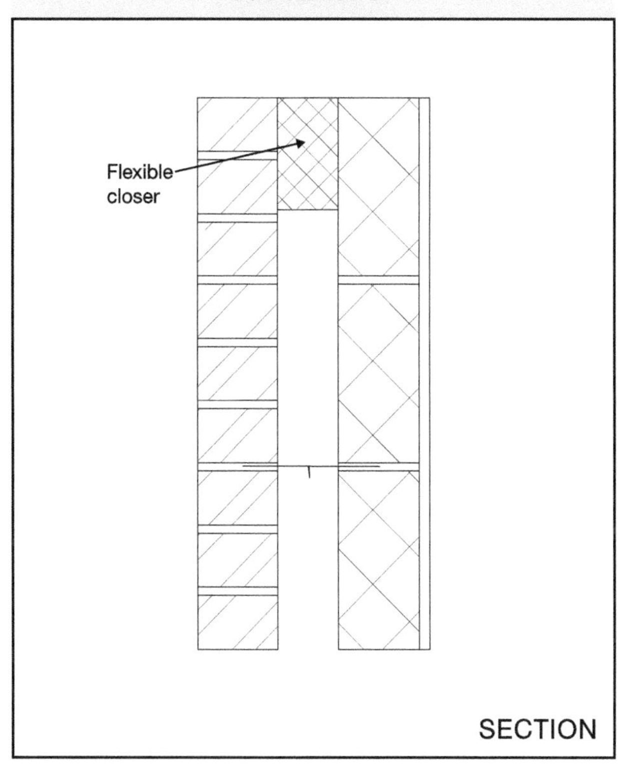

Junctions with separating floors

2.96 There are important details in Section 3 concerning junctions between wall type 2 and separating floors.

Wall type 3: masonry between independent panels

2.97 The resistance to airborne sound depends partly on the type and mass per unit area of the core, and partly on the isolation and mass per unit area of the independent panels.

Note: Wall type 3 can give high resistance to the transmission of both airborne sound and impact sound on the wall.

Construction

2.98 Three wall type 3 constructions (types 3.1, 3.2 and 3.3) are described in this guidance.

2.99 The construction consists of either a solid or cavity masonry core wall with independent panels on both sides. These panels and any frame should not be in contact with the core wall.

2.100 Details of how junctions should be made to limit flanking transmission are also described in this guidance.

2.101 Points to watch

Do

a. Do fill and seal all masonry joints with mortar.
b. Do control flanking transmission from walls and floors connected to the separating wall as described in the guidance on junctions.
c. Do fix the panels or the supporting frames to the ceiling and floor only.
d. Do tape and seal all joints.
e. Do ensure that flue blocks will not adversely affect the sound insulation and that a suitable finish is used over the flue blocks (see BS 1289-1:1986 and seek manufacturer's advice).

Do not

Do not fix, tie or connect the free standing panels or the frame to the masonry core.

Wall ties in cavity masonry cores

2.102 The wall ties used to connect the leaves of a cavity masonry core should be tie type A.

Cavity widths in separating cavity masonry cores

2.103 Recommended cavity widths are minimum values.

2.104 Independent panels.

These panels should meet the following specification:

- minimum mass per unit area of panel (excluding any supporting framework) 20kg/m²;

- panels should consist of either
  a. at least 2 layers of plasterboard with staggered joints, or
  b. a composite panel consisting of 2 sheets of plasterboard separated by a cellular core;
- if the panels are not supported on a frame they should be at least 35mm from the masonry core;
- if the panels are supported on a frame there should be a gap of at least 10mm between the frame and the masonry core.

2.105 Wall type 3.1 Solid masonry core (dense aggregate concrete block), independent panels on both room faces (see Diagrams 2.27 and 2.28)

- minimum mass per unit area of core 300kg/m²;
- minimum core width is determined by structural requirements (see Building Regulation Part A – Structure);
- independent panels on both room faces.

Example of wall type 3.1

The required mass per unit area would be achieved by using

- 140mm block core
- block density 2200kg/m³
- 110mm coursing
- independent panels, each panel of mass per unit area 20kg/m², to be two sheets of plasterboard with joints staggered.

This is an example only. See Annex A for a simplified method of calculating mass per unit area. Alternatively use manufacturer's actual figures where these are available.

Diagram 2.27 **Wall type 3.1 with independent composite panels**

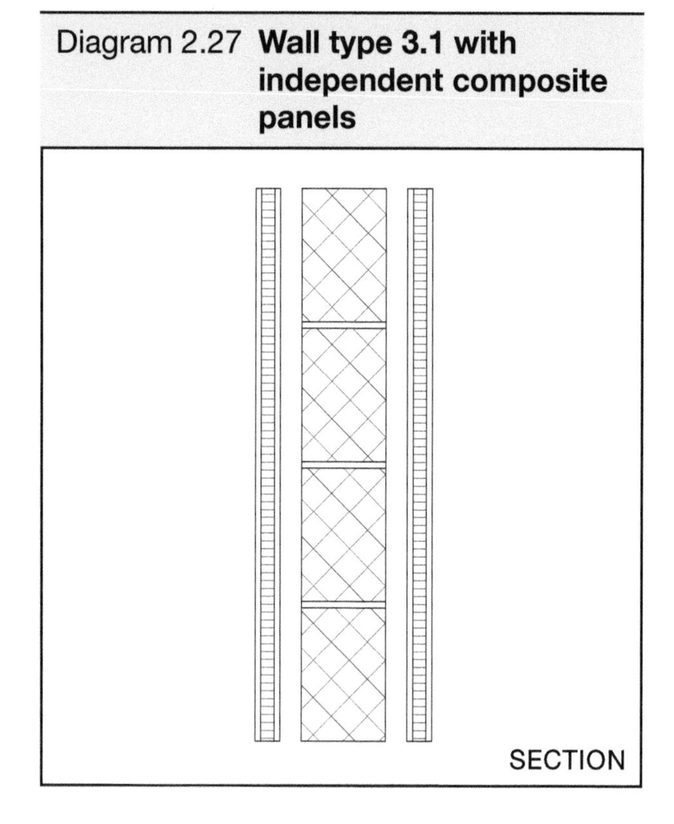

Diagram 2.28 **Wall type 3.1 with independent plasterboard panels**

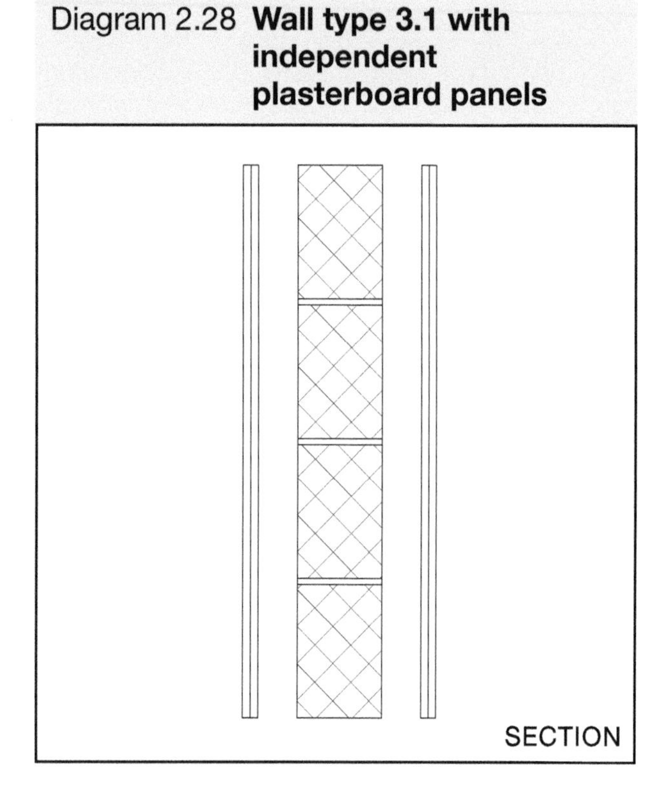

2.106 Wall type 3.2 *Solid masonry core (lightweight concrete block), independent panels on both room faces (see Diagram 2.29)*

- minimum mass per unit area of core 150kg/m²;
- minimum core width is determined by structural requirements (see Building Regulation Part A – Structure);
- independent panels on both room faces.

Example of wall type 3.2

The required mass per unit area would be achieved by using

- 140mm lightweight block core
- block density 1400kg/m³
- 225mm coursing
- independent panels, each panel of mass per unit area 20kg/m², to be two sheets of plasterboard joined by a cellular core

This is an example only. See Annex A for a simplified method of calculating mass per unit area. Alternatively use manufacturer's actual figures where these are vailable.

Diagram 2.29 **Wall type 3.2 with independent composite panels**

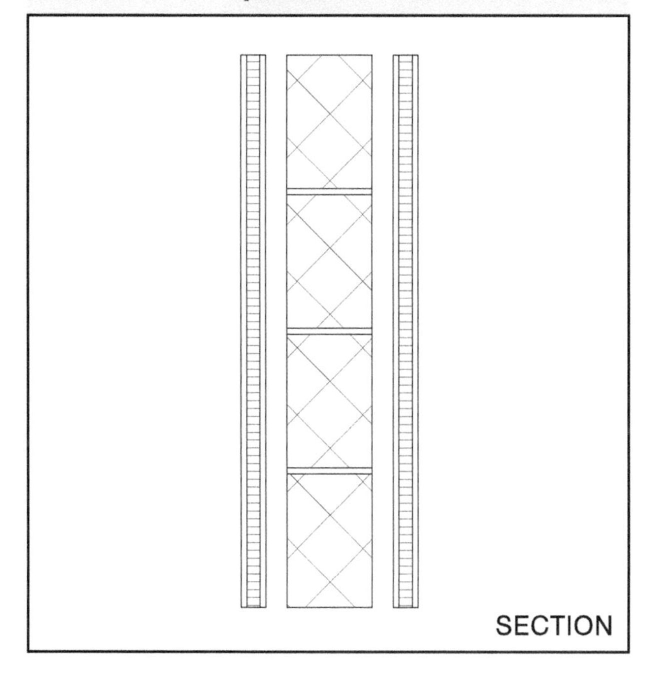

2.107 Wall type 3.3 *Cavity masonry core (brickwork or blockwork), 50mm cavity, independent panels on both room faces (see Diagram 2.30)*

- the core can be of any mass per unit area;
- minimum cavity width of 50mm;
- minimum core width is determined by structural requirements (see Building Regulation Part A – Structure);
- independent panels on both room faces.

Example of wall type 3.3

- two leaves of concrete block
- each leaf at least 100mm thick
- minimum cavity width of 50mm
- independent panels, each panel of mass per unit area 20kg/m², to be two sheets of plasterboard joined by a cellular core

Diagram 2.30 **Wall type 3.3 with independent composite panels**

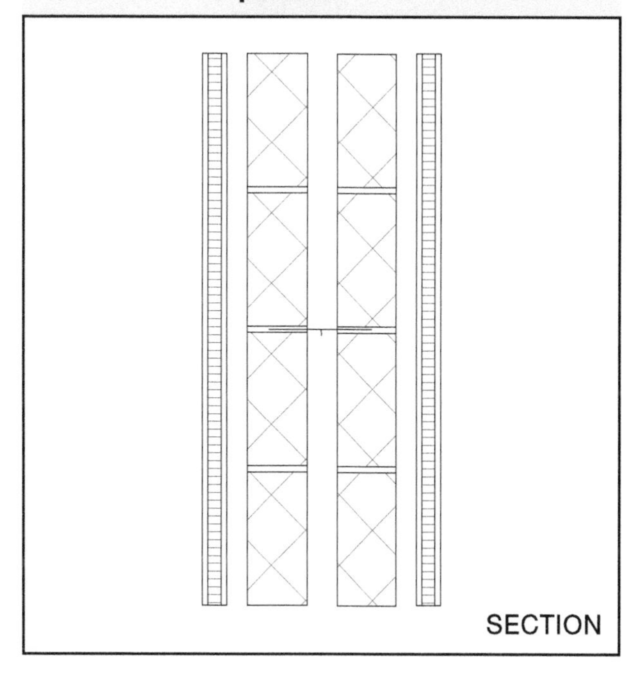

Junction requirements for wall type 3

Junctions with an external cavity wall with masonry inner leaf

2.108 Where the external wall is a cavity wall:

a. the outer leaf of the wall may be of any construction; and

b. the cavity should be stopped with a flexible closer (see Diagram 2.31) unless the cavity is fully filled with mineral wool or expanded polystyrene beads (seek manufacturer's advice for other suitable materials).

2.109 Where the inner leaf of an external cavity wall is masonry:

a. the inner leaf of the external wall should be bonded or tied to the masonry core;

b. the inner leaf of the external wall should be lined with independent panels in the same manner as the separating walls. See Diagram 2.31.

2.110 Where there is a separating floor the masonry inner leaf of the external wall should have a minimum mass per unit area of at least 120kg/m^2 excluding finish.

2.111 Where there is no separating floor and the masonry inner leaf of the external wall is lined with independent panels in the same manner as the separating walls, there is no minimum mass requirement on the masonry inner leaf.

2.112 Where there is no separating floor with separating wall type 3.1 or 3.3, and the masonry inner leaf of the external wall has a mass of at least 120kg/m^2 excluding finish, then the inner leaf of the external wall may be finished with plaster or plasterboard of minimum mass per unit area 10kg/m^2.

Diagram 2.31 **Wall type 3 – external cavity wall with masonry inner leaf**

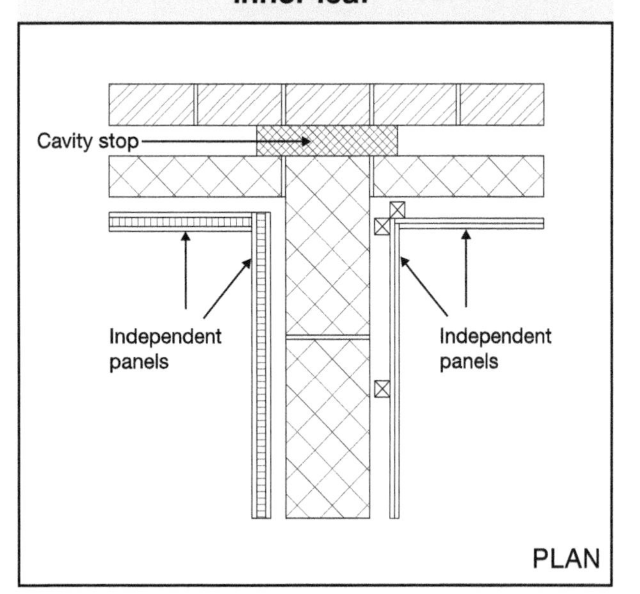

Junctions with an external cavity wall with timber frame inner leaf

2.113 No guidance available (seek specialist advice).

Junctions with an external solid masonry wall

2.114 No guidance available (seek specialist advice).

Junctions with internal framed walls

2.115 Load-bearing framed internal walls should be fixed to the masonry core through a continuous pad of mineral wool. See Diagram 2.32.

2.116 Non-load-bearing internal walls should be butted to the independent panels.

2.117 All joints between internal walls and panels should be sealed with tape or caulked with sealant.

Diagram 2.32 **Wall type 3 – external cavity wall with internal timber wall**

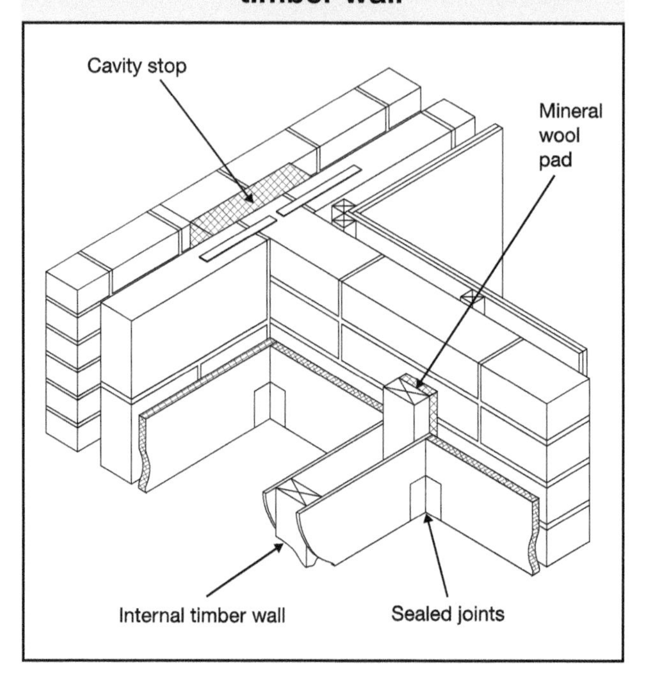

Junctions with internal masonry walls

2.118 Internal walls that abut a type 3 separating wall should not be of masonry construction.

Junctions with internal timber floors

2.119 If the floor joists are to be supported on the separating wall then they should be supported on hangers and should not be built in. See Diagram 2.33.

2.120 Spaces between the floor joists should be sealed with full depth timber blocking.

Junctions with internal concrete floors

Wall types 3.1 and 3.2 (solid masonry core)
2.121 An internal concrete floor slab may only be carried through a solid masonry core if the floor base has a mass per unit area of at least 365kg/m². See Diagram 2-34.

Wall type 3.3 (cavity masonry core)
2.122 Internal concrete floors should generally be built into a cavity masonry core and carried through to the cavity face of the leaf. The cavity should not be bridged.

Diagram 2.33 **Wall type 3 – internal timber floor**

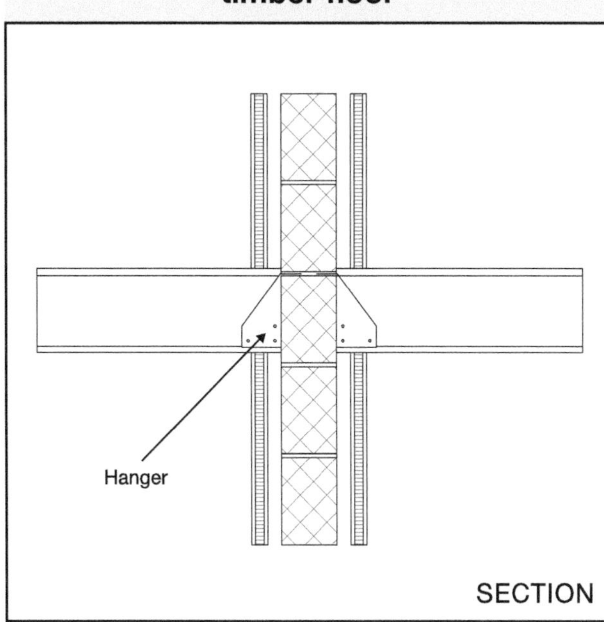

Diagram 2.34 **Wall types 3.1 and 3.2 – internal concrete floor**

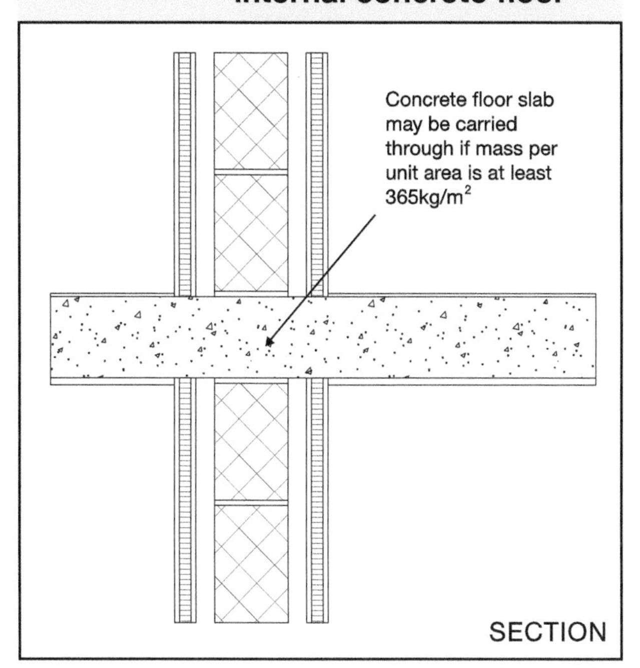

Junctions with timber ground floors

2.123 If the floor joists are to be supported on the separating wall then they should be supported on hangers and should not be built in.

2.124 Spaces between the floor joists should be sealed with full depth timber blocking.

2.125 See Building Regulation Part C – Site preparation and resistance to moisture, and Building Regulation Part L – Conservation of fuel and power.

Junctions with concrete ground floors

2.126 The ground floor may be a solid slab, laid on the ground, or a suspended concrete floor.

Wall type 3.1 and 3.2 (solid masonry core)
2.127 A concrete slab floor on the ground may be continuous under the solid masonry core of a type 3.1 or 3.2 separating wall.

2.128 A suspended concrete floor may only pass under the solid masonry core of a type 3.1 or 3.2 separating wall if the floor has a mass per unit area of at least 365kg/m².

2.129 Hollow core concrete plank and concrete beams with infilling block floors should not be continuous under the solid masonry core of a type 3.1 or 3.2 separating wall.

Wall type 3.3 (cavity masonry core)
2.130 A concrete slab floor on the ground should not be continuous under the cavity masonry core of a type 3.3 separating wall.

2.131 A suspended concrete floor should not be continuous under the cavity masonry core of a type 3.3 separating wall and should be carried through to the cavity face of the leaf. The cavity should not be bridged.

2.132 See Building Regulation Part C – Site preparation and resistance to moisture, and Building Regulation Part L – Conservation of fuel and power.

Diagram 2.35 **Wall types 3.1 and 3.2 – ceiling and roof junction**

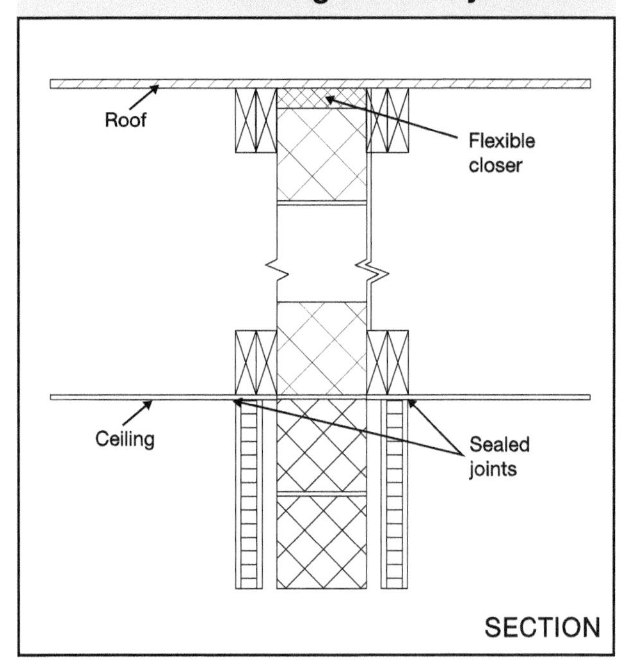

Diagram 2.36 **External cavity wall at eaves level**

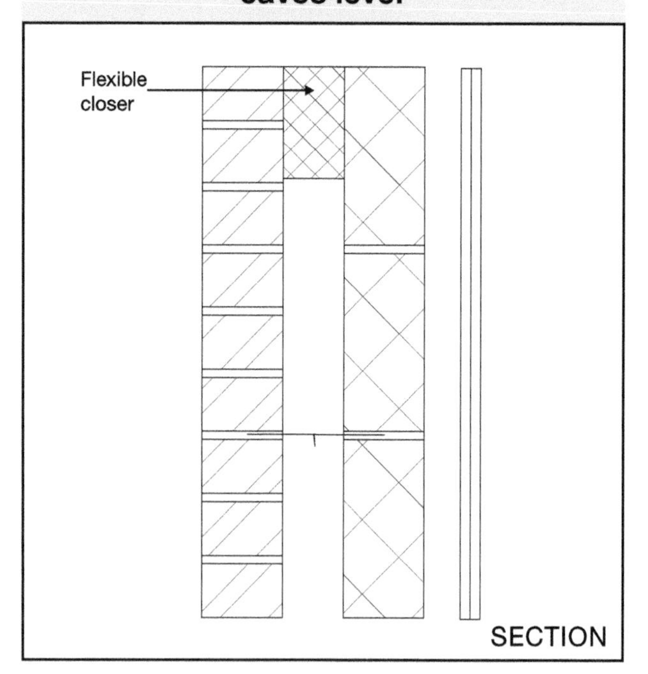

Junctions with ceiling and roof space

2.133 The masonry core should be continuous to the underside of the roof.

2.134 The junction between the separating wall and the roof should be filled with a flexible closer which is also suitable as a fire stop. See Diagram 2.35.

2.135 The junction between the ceiling and independent panels should be sealed with tape or caulked with sealant.

2.136 Where there is an external cavity wall, the cavity should be closed at eaves level with a suitable flexible material (e.g. mineral wool). See Diagram 2.36.

Note: A rigid connection between the inner and external wall leaves should be avoided. If a rigid material is used, then it should only be rigidly bonded to one leaf.

Wall types 3.1 and 3.2 (solid masonry core)
2.137 Where the roof or loft space is not a habitable room and there is a ceiling with a minimum mass per unit area 10kg/m² and with sealed joints, the independent panels may be omitted in the roof space and the mass per unit area of the separating wall above the ceiling may be a minimum of 150kg/m². See Diagram 2.35.

2.138 If lightweight aggregate blocks of density less than 1200kg/m³ are used above ceiling level, then one side should be sealed with cement paint or plaster skim.

Wall type 3.3 (cavity masonry core)
2.139 Where the roof or loft space is not a habitable room and there is a ceiling with a minimum mass per unit area 10kg/m² and with sealed joints, the independent panels may be omitted in the roof space but the cavity masonry core should be maintained to the underside of the roof.

Junctions with separating floors

2.140 There are important details in Section 3 concerning junctions between wall type 3 and separating floors.

Wall type 4: framed walls with absorbent material

2.141 In this guidance only a timber framed wall is described. For steel framed walls, seek advice from the manufacturer.

2.142 The resistance to airborne sound depends on the mass per unit area of the leaves, the isolation of the frames, and the absorption in the cavity between the frames.

Construction

2.143 The construction consists of timber frames, with plasterboard linings on room surfaces and with absorbent material between the frames.

2.144 One wall type 4 construction (type 4.1) is described in this guidance.

2.145 Details of how junctions should be made to limit flanking transmission are also described in this guidance.

2.146 Points to watch

Do

a. Do ensure that where fire stops are needed in the cavity between frames they are either flexible or fixed to only one frame.

b. Do stagger the position of sockets on opposite sides of the separating wall, and use a similar thickness of cladding behind the socket box.

c. Do ensure that each layer of plasterboard is independently fixed to the stud frame.

d. Do control flanking transmission from walls and floors connected to the separating wall as described in the guidance on junctions.

Do not

a. Where it is necessary to connect the two leaves together for structural reasons, do not use ties of greater cross section than 40mm x 3mm fixed to the studwork at or just below ceiling level and do not set them at closer than 1.2m centres.

b. Do not locate sockets back to back. A minimum edge to edge stagger of 150mm is recommended. Do not chase plasterboard.

2.147 Wall type 4.1 *Double leaf frames with absorbent material (see Diagram 2.37)*

- minimum distance between inside lining faces of 200mm;
- plywood sheathing may be used in the cavity as necessary for structural reasons;
- each lining to be two or more layers of plasterboard, each sheet of minimum mass per unit area 10kg/m^2, with staggered joints;
- absorbent material to be unfaced mineral wool batts or quilt (which may be wire reinforced), minimum density 10kg/m^3;
- minimum thickness of absorbent material:

  a. 25mm if suspended in the cavity between frames,

  b. 50mm if fixed to one frame,

  c. 25mm per batt (or quilt) if one is fixed to each frame.

Note: A masonry core may be used where required for structural purposes, but the core should be connected to only one frame.

Diagram 2.37 **Wall type 4.1**

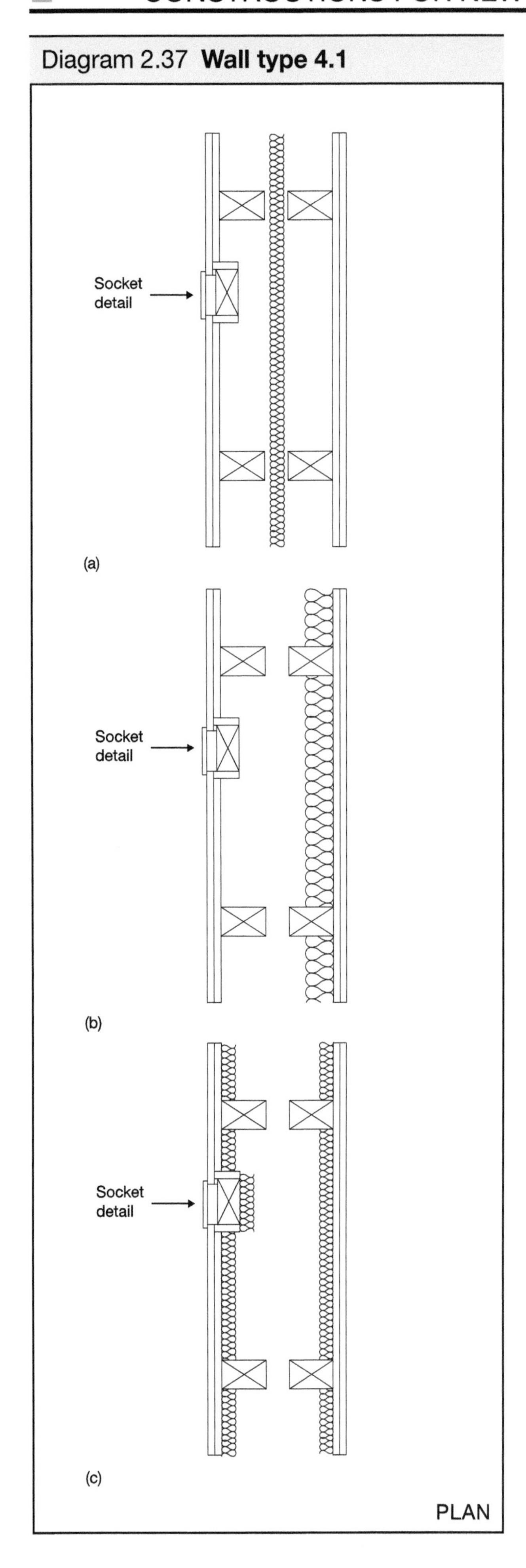

Junction requirements for wall type 4

Junctions with an external cavity wall with masonry inner leaf

2.148 No guidance available (seek specialist advice).

Junctions with an external cavity wall with timber frame inner leaf

2.149 Where the external wall is a cavity wall:

a. the outer leaf of the wall may be of any construction; and

b. the cavity should be stopped between the ends of the separating wall and the outer leaf with a flexible closer. See Diagram 2.38.

2.150 The wall finish of the inner leaf of the external wall should be:

a. one layer of plasterboard; or

b. two layers of plasterboard where there is a separating floor;

c. each sheet of plasterboard of minimum mass per unit area 10kg/m²; and

d. all joints should be sealed with tape or caulked with sealant.

Diagram 2.38 **Wall type 4 – external cavity wall with timber frame inner leaf**

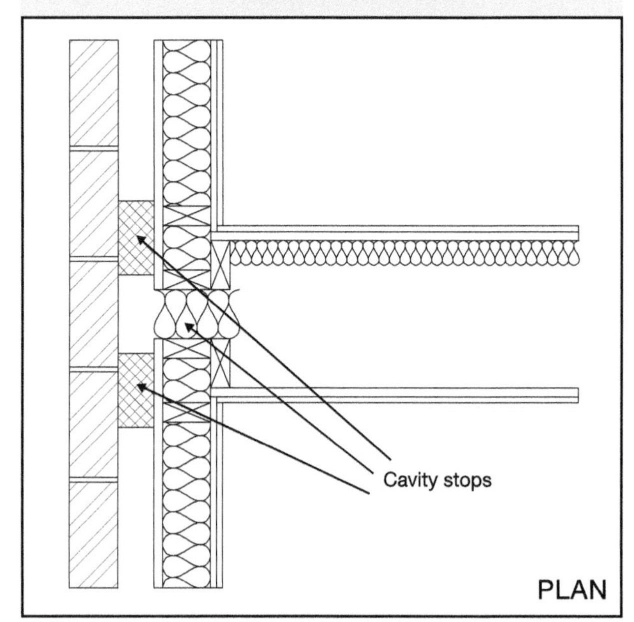

Junctions with an external solid masonry wall

2.151 No guidance available (seek specialist advice).

Junctions with internal framed walls

2.152 There are no restrictions on internal framed walls meeting a type 4 separating wall.

Junctions with internal masonry walls

2.153 There are no restrictions on internal masonry walls meeting a type 4 separating wall.

Junctions with internal timber floors

2.154 Block the air paths through the wall into the cavity by using solid timber blockings or continuous ring beam or joists.

Junctions with internal concrete floors

2.155 No guidance available (seek specialist advice).

Junctions with timber ground floors

2.156 Block the air paths through the wall into the cavity by using solid timber blockings or a continuous ring beam or joists.

2.157 See Building Regulation Part C – Site preparation and resistance to moisture, and Building Regulation Part L – Conservation of fuel and power.

Junctions with concrete ground floors

2.158 The ground floor may be a solid slab, laid on the ground, or a suspended concrete floor. A concrete slab floor on the ground may be continuous under a type 4 separating wall. A suspended concrete floor may only pass under a wall type 4 if the floor has a mass per unit area of at least 365kg/m^2.

2.159 See Building Regulation Part C – Site preparation and resistance to moisture, and Building Regulation Part L – Conservation of fuel and power.

Junctions with ceiling and roof space

2.160 The wall should preferably be continuous to the underside of the roof.

2.161 The junction between the separating wall and the roof should be filled with a flexible closer.

2.162 The junction between the ceiling and the wall linings should be sealed with tape or caulked with sealant.

Where the roof or loft space is not a habitable room and there is a ceiling with a minimum mass per unit area 10kg/m^2 and with sealed joints, either:

a. the linings on each frame may be reduced to two layers of plasterboard, each sheet of minimum mass per unit area 10kg/m^2; or

b. the cavity may be closed at ceiling level without connecting the two frames rigidly together and then one frame may be used in the roof space provided there is a lining of two layers of plasterboard, each sheet of minimum mass per unit area 10kg/m^2, on both sides of the frame.

2.163 Where there is an external wall cavity, the cavity should be closed at eaves level with a suitable material.

Junctions with separating floors

2.164 There are important details in Section 3 concerning junctions between wall type 4 and separating floors.

Section 3: Separating floors and associated flanking constructions for new buildings

Introduction

3.1 This Section gives examples of floor types which, if built correctly, should achieve the performance standards set out in Section 0: Performance – Table 1a.

3.2 The guidance in this section is not exhaustive and other designs, materials or products may be used to achieve the performance standards set out in Section 0: Performance – Table 1a. Advice should be sought from the manufacturer or other appropriate source.

3.3 The floors are grouped into three main types. See Diagram 3.1.

3.4 Floor type 1: *Concrete base with ceiling and soft floor covering*

The resistance to airborne sound depends mainly on the mass per unit area of the concrete base and partly on the mass per unit area of the ceiling. The soft floor covering reduces impact sound at source.

3.5 Floor type 2: *Concrete base with ceiling and floating floor*

The resistance to airborne and impact sound depends on the mass per unit area of the concrete base, as well as the mass per unit area and isolation of the floating layer and the ceiling. The floating floor reduces impact sound at source.

3.6 Floor type 2: *Floating floor*

Floor type 2 requires one of the floating floors described in this section. The description of floor type 2 contains a suffix (a), (b) or (c) which refers to the floating floor used.

3.7 Floor type 3: *Timber frame base with ceiling and platform floor*

The resistance to airborne and impact sound depends on the structural floor base and the isolation of the platform floor and the ceiling. The platform floor reduces impact sound at source.

3.8 Ceiling treatment

Each floor type requires one of the ceiling treatments described in this section. The description of each floor type contains a suffix A, B or C that refers to the ceiling treatment used.

3.9 Within each floor type the constructions are ranked, as far as possible, with constructions providing better sound insulation given first.

Junctions between separating floors and other building elements

3.10 In order for the floor construction to be fully effective, care should be taken to correctly detail the junctions between the separating floor and other elements such as external walls, separating walls and floor penetrations. Recommendations are also given for the construction of these other elements where it is necessary to control flanking transmission. Notes and diagrams explain the junction details for each of the separating floor types.

Diagram 3.1 **Types of separating floor**

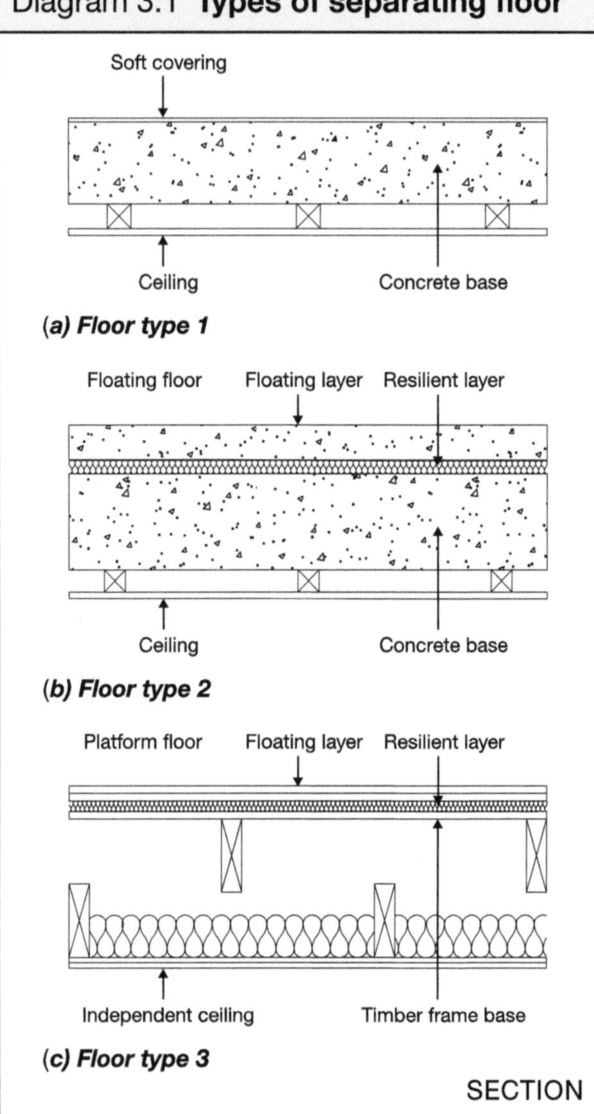

3.11 Table 3.1 indicates the inclusion of guidance in this document on the junctions that may occur between each of the separating floor types and various attached building elements.

Table 3.1 **Separating floor junctions reference table**

	Separating floor type		
Building element attached to separating wall	**Type 1**	**Type 2**	**Type 3**
External cavity wall with masonry inner leaf	G	G	G
External cavity wall with timber frame inner leaf	G	G	G
External solid masonry wall	N	N	N
Internal wall – framed	G	G	G
Internal wall – masonry	G	G	N
Floor penetrations	G	G	G
For flats the following may also apply:			
Separating wall type 1 – solid masonry	G	G	G
Separating wall type 2 – cavity masonry	G	G	G
Separating wall type 3 – masonry between independent panels	G	G	G
Separating wall type 4 – framed wall with absorbent material	N	N	G

Key: G = guidance available; N = no guidance available (seek specialist advice)

Note:

Where any building element functions as a separating element (e.g. a ground floor that is also a separating floor for a basement flat) then the separating element requirements should take precedence.

Beam and block floors

3.12 For beam and block separating floors, seek advice from the manufacturer.

Mass per unit area of floors

3.13 The mass per unit area of a floor is expressed in kilograms per square metre (kg/m²). The mass per unit area of floors should be obtained from manufacturer's data or calculated using the method shown in Annex A.

3.14 The density of the materials used (and on which the mass per unit area of the floor depends) is expressed in kilograms per cubic metre (kg/m³).

3.15 Where appropriate, the mass per unit area of a bonded screed may be included in the calculation of the mass per unit area of the floor.

3.16 The mass per unit area of a floating screed should not be included in the calculation of the mass per unit area of the floor.

Ceiling treatments

3.17 Each floor type should use one of the following three ceiling treatments (A, B or C). See Diagram 3.2.

3.18 The ceiling treatments are ranked, in order of sound insulation performance from A to C, with constructions providing higher sound insulation given first.

Note: Use of a better performing ceiling than that described in the guidance should improve the sound insulation of the floor provided there is no significant flanking transmission.

3.19 Ceiling treatment A, *independent ceiling with absorbent material*

Ceiling treatment A should meet the following specification:

- at least 2 layers of plasterboard with staggered joints;
- minimum total mass per unit area of plasterboard 20kg/m²;
- an absorbent layer of mineral wool (minimum thickness 100mm, minimum density 10kg/m³) laid in the cavity formed above the ceiling.

The ceiling should be supported by one of the following methods:

- **Floor types 1, 2 and 3.** Use independent joists fixed only to the surrounding walls. A clearance of at least 100mm should be left between the top of the plasterboard forming the ceiling and the underside of the base floor.
- **Floor type 3.** Use independent joists fixed to the surrounding walls with additional support provided by resilient hangers attached directly to the floor. A clearance of at least 100mm should be left between the top of the ceiling joists and the underside of the base floor.

3.20 Points to watch:

Do

Do seal the perimeter of the independent ceiling with tape or sealant.

Do not

Do not create a rigid or direct connection between the independent ceiling and the floor base.

3.21 Ceiling treatment B, *plasterboard on proprietary resilient bars with absorbent material*

Ceiling treatment B should meet the following specification:

- single layer of plasterboard, minimum mass per unit area of plasterboard 10kg/m²;
- fixed using proprietary resilient metal bars. On concrete floors, these resilient metal bars should be fixed to timber battens. For fixing details, seek advice from the manufacturer;
- an absorbent layer of mineral wool (minimum density 10kg/m³) that fills the ceiling void.

3.22 Ceiling treatment C, *plasterboard on timber battens or proprietary resilient channels with absorbent material*

Ceiling treatment C should meet the following specification:

- single layer of plasterboard, minimum mass per unit area 10kg/m²;
- fixed using timber battens or proprietary resilient channels;
- if resilient channels are used, incorporate an absorbent layer of mineral wool minimum density 10kg/m³) that fills the ceiling void.

Note: Electrical cables give off heat when in use and special precautions may be required when they are covered by thermally insulating materials. See BRE BR 262, Thermal Insulation: avoiding risks, section 2.4. Installing recessed light fittings in ceiling treatments A to C can reduce their resistance to the passage of airborne and impact sound.

Diagram 3.2 **Ceiling treatments A, B and C**

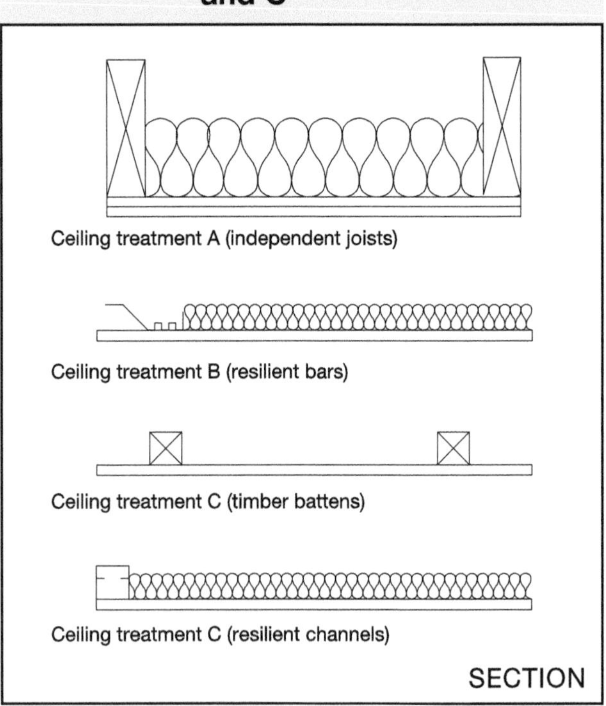

Floor type 1: concrete base with ceiling and soft floor covering

3.23 The resistance to airborne sound depends mainly on the mass per unit area of the concrete base and partly on the mass per unit area of the ceiling. The soft floor covering reduces impact sound at source.

Constructions

3.24 The construction consists of a concrete floor base with a soft floor covering and a ceiling.

3.25 Two floor type 1 constructions (types 1.1C and 1.2B) are described in this guidance which should be combined with the appropriate ceiling and soft floor covering.

3.26 Details of how junctions should be made to limit flanking transmission are also described in this guidance.

3.27 Points to watch

Do

a. Do fix or glue the soft floor covering to the floor. (N.B. allow for future replacement.)

b. Do fill all joints between parts of the floor to avoid air paths.

c. Do give special attention to workmanship and detailing at the perimeter and wherever a pipe or duct penetrates the floor in order to reduce flanking transmission and to avoid air paths.

d. Do build a separating concrete floor into the walls around its entire perimeter where the walls are masonry.

e. Do fill with mortar any gap that may form between the head of a masonry wall and the underside of the concrete floor.

f. Do control flanking transmission from walls connected to the separating floor as described in the guidance on junctions.

Do not

a. Do not allow the floor base to bridge a cavity in a cavity masonry wall.

b. Do not use non-resilient floor finishes that are rigidly connected to the floor base.

3.28 Soft floor covering

The soft floor covering should meet the following specification:

- any resilient material, or material with a resilient base, with an overall uncompressed thickness of at least 4.5mm; or
- any floor covering with a weighted reduction in impact sound pressure level (ΔL_w) of not less than 17dB when measured in accordance with BS EN ISO 140-8:1998 and calculated in accordance with BS EN ISO 717-2:1997.

3.29 Floor type 1.1C *Solid concrete slab (cast in situ, with or without permanent shuttering), soft floor covering, ceiling treatment C (see Diagram 3.3)*

- minimum mass per unit area of 365kg/m^2 (including shuttering only if it is solid concrete or metal) and including any bonded screed;
- soft floor covering essential;
- ceiling treatment C (or better) essential.

Diagram 3.3 **Floor type 1.1C – floor type 1.1 with ceiling treatment C**

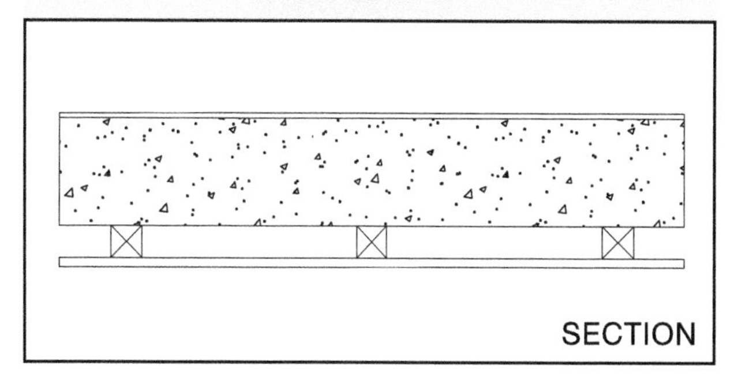

3.30 Floor Type 1.2B *Concrete planks (solid or hollow), soft floor covering, ceiling treatment B (see Diagram 3.4)*

- minimum mass per unit area of planks and any bonded screed of 365kg/m^2;
- use a regulating floor screed;
- all floor joints fully grouted to ensure air tightness;
- soft floor covering essential;
- ceiling treatment B (or better) essential.

Diagram 3.4 **Floor type 1.2B – floor type 1.2 with ceiling treatment B**

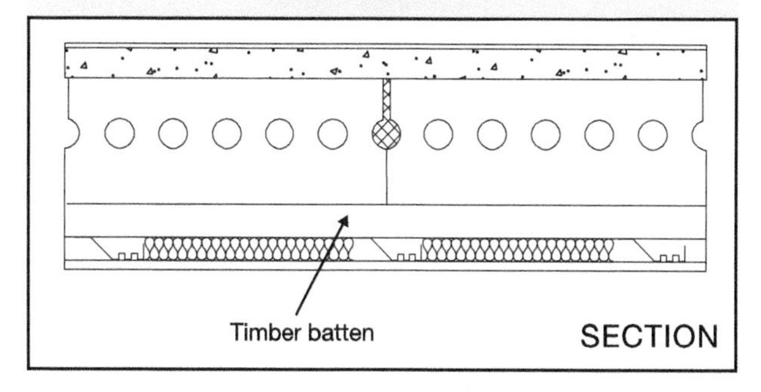

Junction requirements for floor type 1

Junctions with an external cavity wall with masonry inner leaf

3.31 Where the external wall is a cavity wall:

a. the outer leaf of the wall may be of any construction; and

b. the cavity should be stopped with a flexible closer (see Diagram 3.5) ensuring adequate drainage, unless the cavity is fully filled with mineral wool or expanded polystyrene beads (seek manufacturer's advice for other suitable materials).

3.32 The masonry inner leaf of an external cavity wall should have a mass per unit area of at least 120kg/m^2 excluding finish.

3.33 The floor base (excluding any screed) should be built into a cavity masonry external wall and carried through to the cavity face of the inner leaf. The cavity should not be bridged.

Floor type 1.2B

3.34 Where floor type 1.2B is used and the planks are parallel to the external wall the first joint should be a minimum of 300mm from the cavity face of the inner leaf. See Diagram 3.5.

3.35 See details in Section 2 concerning the use of wall ties in external masonry cavity walls.

Diagram 3.5 **Floor type 1.2B – external cavity wall with masonry inner leaf**

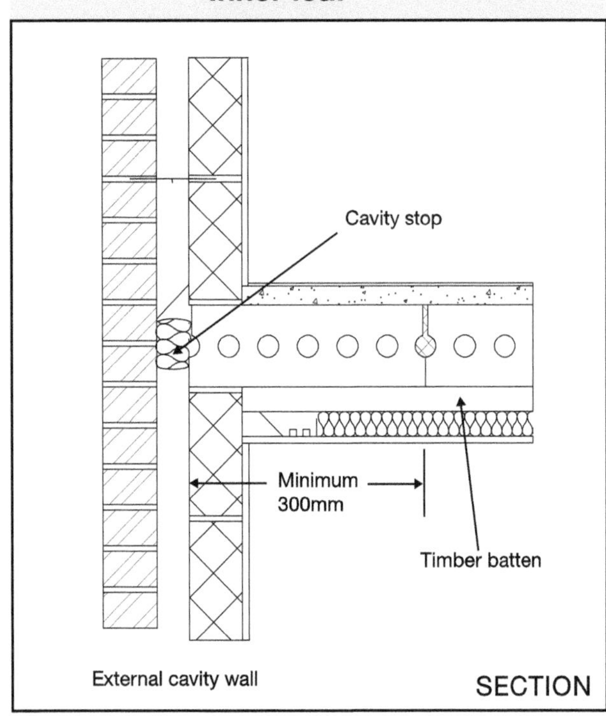

Junctions with an external cavity wall with timber frame inner leaf

3.36 Where the external wall is a cavity wall:

a. the outer leaf of the wall may be of any construction; and

b. the cavity should be stopped with a flexible closer;

c. the wall finish of the inner leaf of the external wall should be two layers of plasterboard, each sheet of plasterboard to be of minimum mass per unit area 10kg/m², and all joints should be sealed with tape or caulked with sealant.

Junctions with an external solid masonry wall

3.37 No guidance available (seek specialist advice).

Junctions with internal framed walls

3.38 There are no restrictions on internal framed walls meeting a type 1 separating floor.

Junctions with internal masonry walls

3.39 The floor base should be continuous through, or above, an internal masonry wall.

3.40 The mass per unit area of any load-bearing internal wall or any internal wall rigidly connected to a separating floor should be at least 120kg/m² excluding finish.

Junctions with floor penetrations (excluding gas pipes)

3.41 Pipes and ducts that penetrate a floor separating habitable rooms in different flats should be enclosed for their full height in each flat. See Diagram 3-6.

3.42 The enclosure should be constructed of material having a mass per unit area of at least 15kg/m². Either line the enclosure or wrap the duct or pipe within the enclosure with 25mm unfaced mineral fibre.

3.43 Penetrations through a separating floor by ducts and pipes should have fire protection to satisfy Building Regulation Part B – Fire safety. Fire stopping should be flexible and prevent rigid contact between the pipe and floor.

Diagram 3.6 **Floor type 1 – floor penetrations**

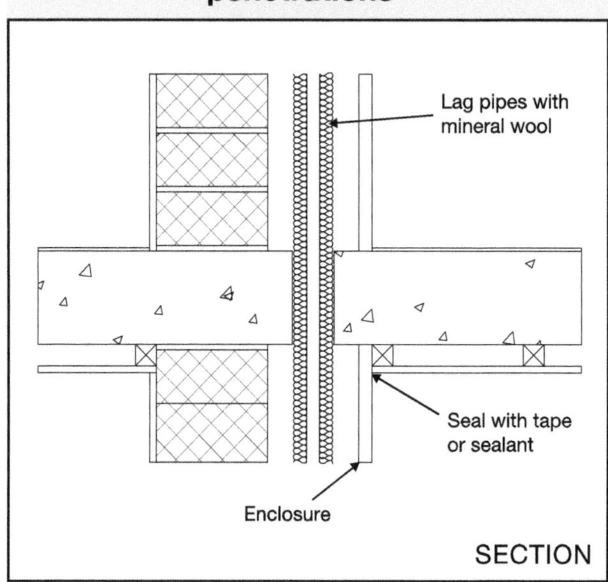

Note: There are requirements for ventilation of ducts at each floor where they contain gas pipes. Gas pipes may be contained in a separate ventilated duct or they can remain unenclosed. Where a gas service is installed, it shall comply with relevant codes and standards to ensure safe and satisfactory operation. See The Gas Safety (Installation and Use) Regulations 1998, SI 1998 No.2451.

For flats where there are separating walls the following may also apply:

Junctions with separating wall type 1 – solid masonry

Floor type 1.1C

Diagram 3.7 **Floor type 1.1C – wall type 1**

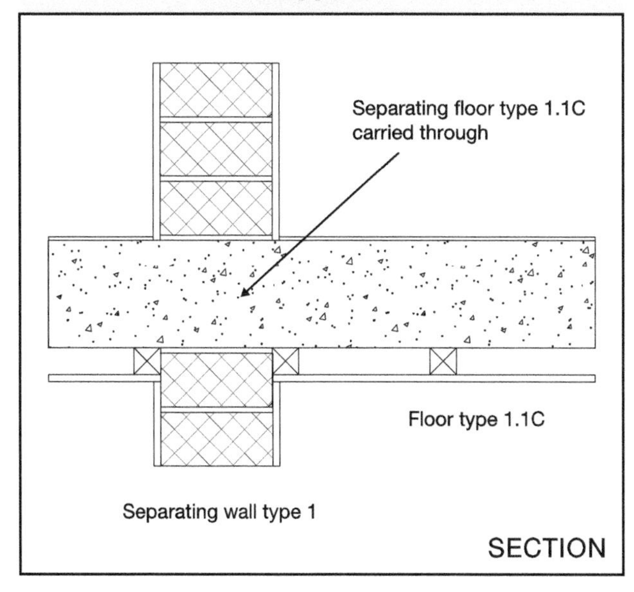

3.44 A separating floor type 1.1C base (excluding any screed) should pass through a separating wall type 1. See Diagram 3.7.

Floor type 1.2B

Diagram 3.8 **Floor type 1.2B – wall type 1**

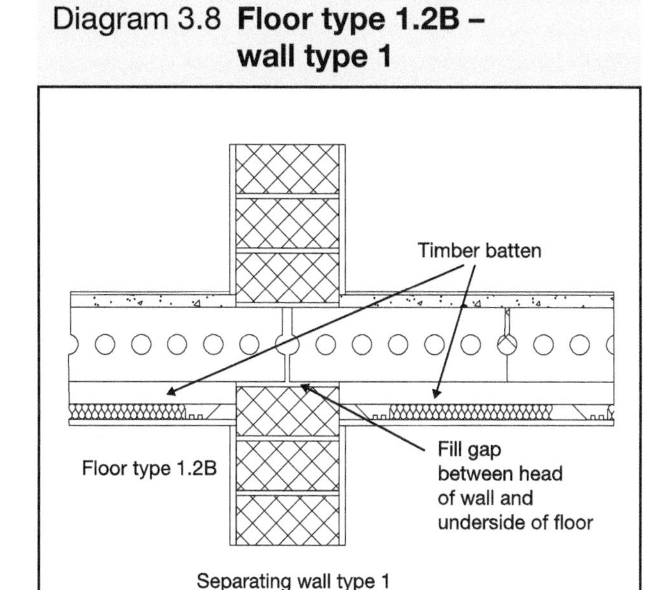

3.45 A separating floor type 1.2B base (excluding any screed) should not be continuous through a separating wall type 1. See Diagram 3.8.

Junctions with separating wall type 2 – cavity masonry

3.46 The mass per unit area of any leaf that is supporting or adjoining the floor should be at least 120kg/m² excluding finish.

3.47 The floor base (excluding any screed) should be carried through to the cavity face of the leaf. The wall cavity should not be bridged. See Diagram 3.9.

Floor type 1.2B

3.48 Where floor type 1.2B is used and the planks are parallel to the separating wall the first joint should be a minimum of 300mm from the inner face of the adjacent cavity leaf. See Diagram 3.9.

Diagram 3.9 **Floor types 1.1C and 1.2B – wall type 2**

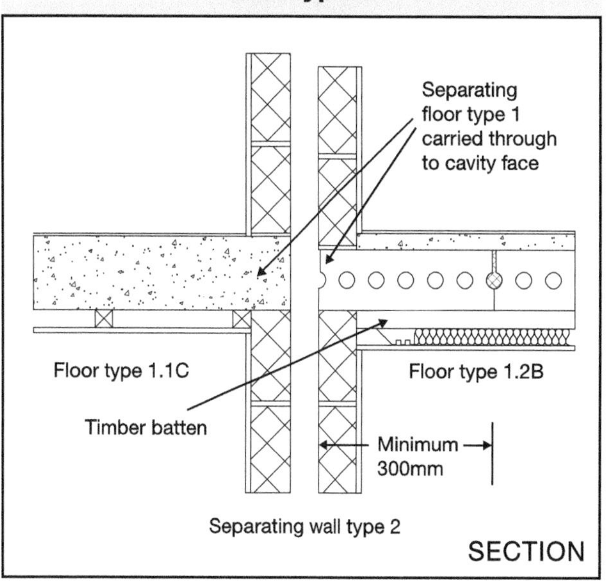

Junctions with separating wall type 3 – masonry between independent panels

Junctions with separating wall type 3.1 and 3.2 (solid masonry core)

Floor type 1.1C

3.49 A separating floor type 1.1C base (excluding any screed) should pass through separating wall types 3.1 and 3.2. See Diagram 3.10.

Diagram 3.10 **Floor type 1.1C – wall types 3.1 and 3.2**

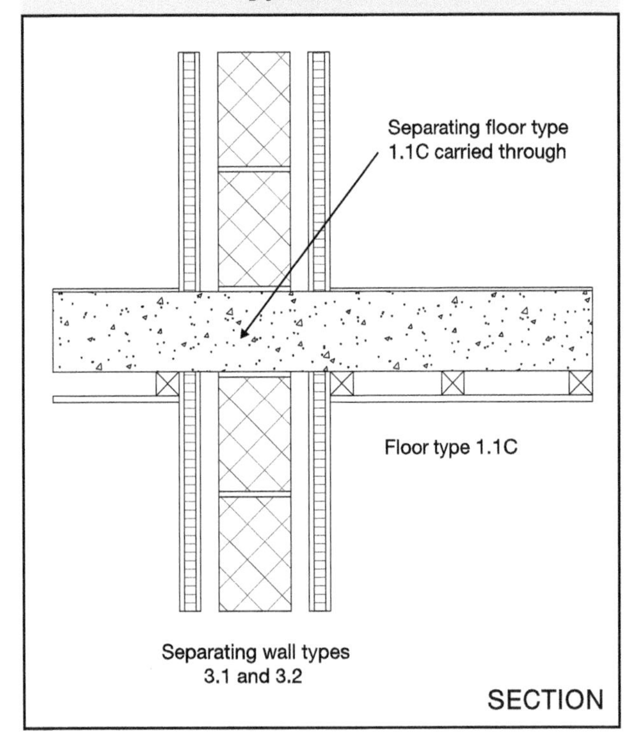

Floor type 1.2B

3.50 A separating floor type 1.2B base (excluding any screed) should not be continuous through a separating wall type 3.

3.51 Where separating wall type 3.2 is used with floor type 1.2B and the planks are parallel to the separating wall the first joint should be a minimum of 300mm from the centreline of the masonry core.

Junctions with separating wall type 3.3 (cavity masonry core)

3.52 The mass per unit area of any leaf that is supporting or adjoining the floor should be at least 120kg/m^2 excluding finish.

3.53 The floor base (excluding any screed) should be carried through to the cavity face of the leaf of the core. The cavity should not be bridged.

Floor type 1.2B

3.54 Where floor type 1.2B is used and the planks are parallel to the separating wall the first joint should be a minimum of 300mm from the inner face of the adjacent cavity leaf of the masonry core.

Junctions with separating wall type 4 – timber frames with absorbent material

3.55 No guidance available (seek specialist advice).

Floor type 2: concrete base with ceiling and floating floor

3.56 The resistance to airborne and impact sound depends on the mass per unit area of the concrete base, as well as the mass per unit area and isolation of the floating layer and the ceiling. The floating floor reduces impact sound at source.

Constructions

3.57 The construction consists of a concrete floor base with a floating floor and a ceiling. The floating floor consists of a floating layer and a resilient layer.

3.58 Two floor type 2 constructions (types 2.1C and 2.2B) are described in this guidance, which should be combined with the appropriate ceiling and any one of the three floating floor options (a), (b) or (c).

3.59 Details of how junctions should be made to limit flanking transmission are also described in this guidance.

Limitations

3.60 Where resistance to airborne sound only is required the full construction should still be used.

3.61 Points to watch

Do

a. Do fill all joints between parts of the floor to avoid air paths.
b. Do give special attention to workmanship and detailing at the perimeter and wherever a pipe or duct penetrates the floor in order to reduce flanking transmission and to avoid air paths.
c. Do build a separating concrete floor base into the walls around its entire perimeter where the walls are masonry.
d. Do fill with mortar any gap that may form between the head of a masonry wall and the underside of the concrete floor.
e. Do control flanking transmission from walls connected to the separating floor as described in the guidance on junctions.

Do not

Do not allow the floor base to bridge a cavity in a cavity masonry wall.

Floating floors (floating layers and resilient layers)

3.62 The floating floor consists of a floating layer and resilient layer. See Diagram 3.11.

3.63 Points to watch

Do

a. Do leave a small gap (as advised by the manufacturer) between the floating layer and wall at all room edges and fill with a flexible sealant.

b. Do leave a small gap (approx. 5mm) between skirting and floating layer and fill with a flexible sealant.

c. Do lay resilient materials in rolls or sheets with lapped joints or with joints tightly butted and taped.

d. Do use paper facing on the upper side of fibrous materials to prevent screed entering the resilient layer.

Do not

a. Do not bridge between the floating layer and the base or surrounding walls (e.g. with services or fixings that penetrate the resilient layer).

b. Do not let the floating screed create a bridge (for example through a gap in the resilient layer) to the concrete floor base or surrounding walls.

Diagram 3.11 **Floating floors (a) and (b)**

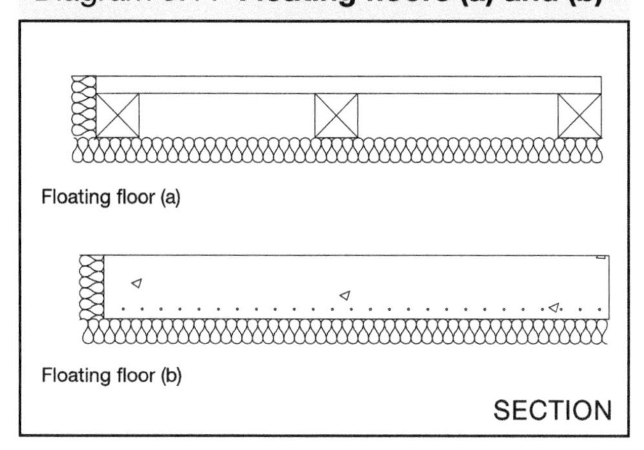

3.64 Floating floor (a) *Timber raft floating layer with resilient layer*

Floating floor (a) should meet the following specification:

- timber raft of board material (with bonded edges, e.g. tongued and grooved) of minimum mass per unit area 12kg/m^2, fixed to 45mm x 45mm battens;
- timber raft to be laid loose on the resilient layer, battens should not be laid along any joints in the resilient layer;
- resilient layer of mineral wool with density 36kg/m^3 and minimum thickness 25mm. The resilient layer may be paper faced on the underside.

3.65 Floating floor (b) *Sand cement screed floating layer with resilient layer*

Floating floor (b) should meet the following specification:

- floating layer of 65mm sand cement screed or a suitable proprietary creed product with a mass per unit area of at least 80kg/m^2. Ensure that the resilient layer is protected while the screed is being laid. A 20–50mm wire mesh may be used for this purpose;
- resilient layer consisting of either:
  a. a layer of mineral wool of minimum thickness 25mm with density 36kg/m^3, paper faced on the upper side to prevent the screed entering the resilient layer, or
  b. an alternative type of resilient layer which meets the following two requirements:
     i. maximum dynamic stiffness (measured according to BS EN 29052-1:1992) of 15MN/m^3, and
     ii. minimum thickness of 5mm under the load specified in the measurement procedure of BS EN 29052-1:1992, 1.8kPa to 2.1kPa.

Note: For proprietary screed products, seek advice from the manufacturer.

3.66 Floating floor (c) *Performance based approach*

Floating floor (c) should meet the following specification:

- rigid boarding above a resilient and/or damping layer(s); with
- weighted reduction in impact sound pressure level (ΔL_w) of not less than 29dB when measured according to BS EN ISO 140-8:1998 and rated according to BS EN ISO 717-2:1997. (See Annex B: Supplementary guidance on acoustic measurement standards.) The performance value ΔL_w should be achieved when the floating floor is both loaded and unloaded as described in BS EN ISO 140-8:1998 for category II systems.

Note: For details on the performance and installation of proprietary floating floors, seek advice from the manufacturer.

3.67 Floor type 2.1C *Solid concrete slab (cast in-situ, with or without permanent shuttering), floating floor, ceiling treatment C (see Diagrams 3.12 and 3.13)*

- minimum mass per unit area of 300kg/m² (including shuttering only if it is solid concrete or metal), and including any bonded screed;
- regulating floor screed optional;
- floating floor (a), (b) or (c) essential;
- ceiling treatment C (or better) essential.

3.68 Floor type 2.2B *Concrete planks (solid or hollow), floating floor, ceiling treatment B (see Diagrams 3.14 and 3.15)*

- minimum mass per unit area of planks and any bonded screed of 300g/m²;
- use a regulating floor screed;
- all floor joints fully grouted to ensure air tightness;
- floating floor (a), (b) or (c) essential;
- ceiling treatment B (or better) essential.

Diagram 3.12 **Floor type 2.1C(a) – floor type 2.1 with ceiling treatment C and floating floor (a)**

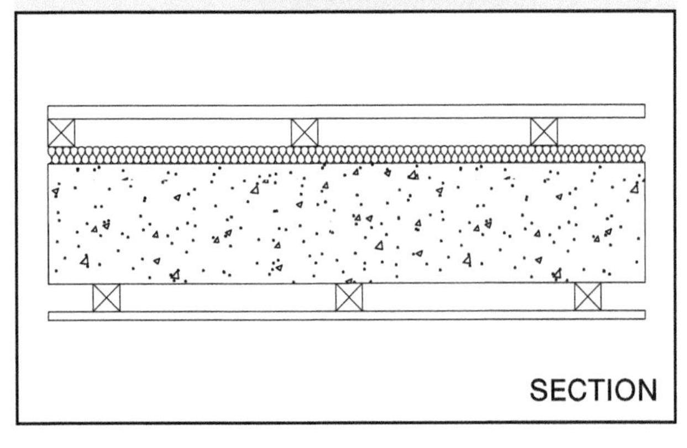

Diagram 3.13 **Floor type 2.1C(b) – floor type 2.1 with ceiling treatment C and floating floor (b)**

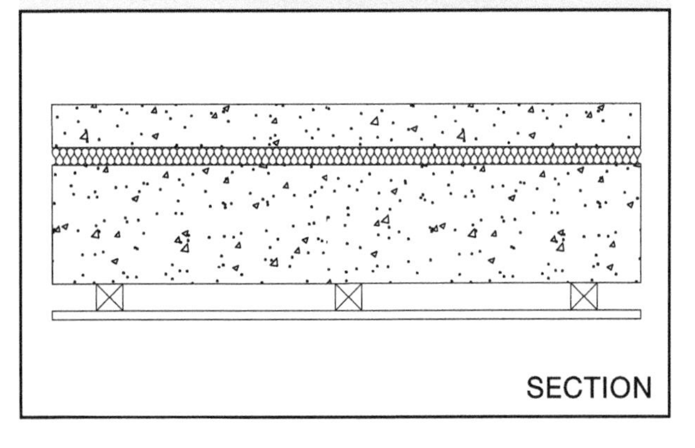

Diagram 3.14 **Floor type 2.2B(a) – floor type 2.2 with ceiling treatment B and floating floor (a)**

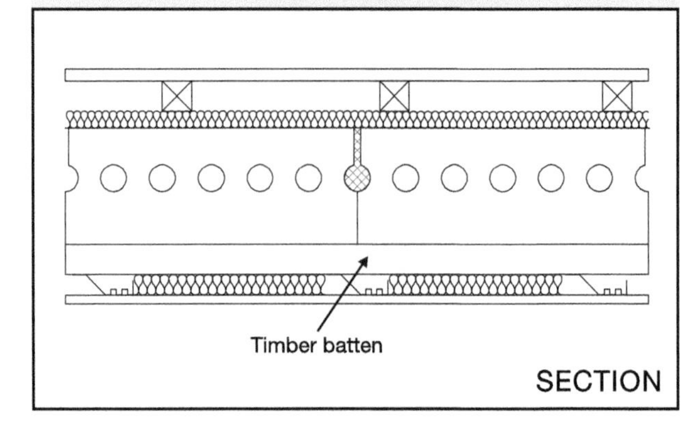

Diagram 3.15 **Floor type 2.2B(b) – floor type 2.2 with ceiling treatment B and floating floor (b)**

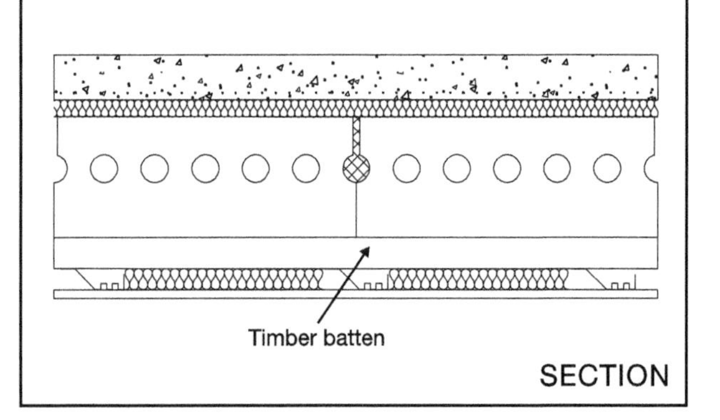

Junction requirements for floor type 2

Junctions with an external cavity wall with masonry inner leaf

3.69 Where the external wall is a cavity wall:

a. the outer leaf of the wall may be of any construction; and

b. the cavity should be stopped with a flexible closer (see Diagram 3.16) ensuring adequate drainage, unless the cavity is fully filled with mineral wool or expanded polystyrene beads (seek manufacturer's advice for other suitable materials).

3.70 The masonry inner leaf of an external cavity wall should have a mass per unit area of at least 120kg/m² excluding finish.

3.71 The floor base (excluding any screed) should be built into a cavity masonry external wall and carried through to the cavity face of the inner leaf. The cavity should not be bridged.

Floor 2.2B

3.72 Where floor 2.2B is used and the planks are parallel to the external wall the first joint should be a minimum of 300mm from the cavity face of the inner leaf. See Diagram 3.16.

3.73 See details in Section 2 concerning the use of wall ties in external masonry cavity walls.

Diagram 3.16 **Floor type 2 – external cavity wall with masonry internal leaf**

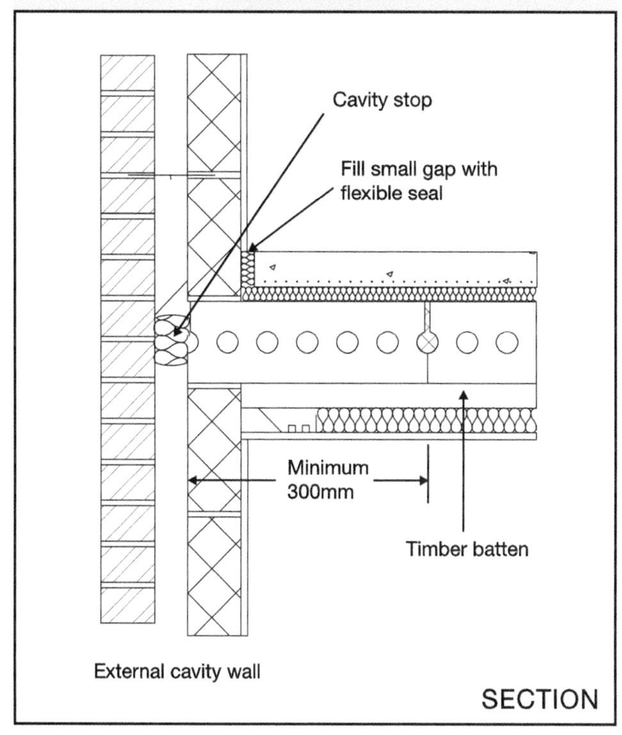

Junctions with an external cavity wall with timber frame inner leaf

3.74 Where the external wall is a cavity wall:

a. the outer leaf of the wall may be of any construction;

b. the cavity should be stopped with a flexible closer; and

c. the wall finish of the inner leaf of the external wall should be two layers of plasterboard, each sheet of plasterboard to be of minimum mass per unit area 10kg/m², and all joints should be sealed with tape or caulked with sealant.

Junctions with an external solid masonry wall

3.75 No guidance available (seek specialist advice).

Junctions with internal framed walls

3.76 There are no restrictions on internal framed walls meeting a type 2 separating floor.

Junctions with internal masonry walls

3.77 The floor base should be continuous through, or above an internal masonry wall.

3.78 The mass per unit area of any load-bearing internal wall or any internal wall rigidly connected to a separating floor should be at least 120kg/m² excluding finish.

Junctions with floor penetrations (excluding gas pipes)

3.79 Pipes and ducts that penetrate a floor separating habitable rooms in different flats should be enclosed for their full height in each flat. See Diagram 3.17.

3.80 The enclosure should be constructed of material having a mass per unit area of at least 15kg/m². Either line the enclosure, or wrap the duct or pipe within the enclosure, with 25mm unfaced mineral wool.

3.81 Leave a small gap (approx. 5mm) between the enclosure and floating layer and seal with sealant or neoprene. Where floating floor (a) or (b) is used the enclosure may go down to the floor base, but ensure that the enclosure is isolated from the floating layer.

3.82 Penetrations through a separating floor by ducts and pipes should have fire protection to satisfy Building Regulation Part B – Fire safety. Fire stopping should be flexible and also prevent rigid contact between the pipe and floor.

Note: There are requirements for ventilation of ducts at each floor where they contain gas pipes. Gas pipes may be contained in a separate ventilated duct or they can remain unenclosed. Where a gas service is installed, it shall comply with relevant codes and standards to ensure safe and satisfactory operation. See The Gas Safety (Installation and Use) Regulations 1998, SI 1998/2451.

Diagram 3.17 Floor type 2 – floor penetrations

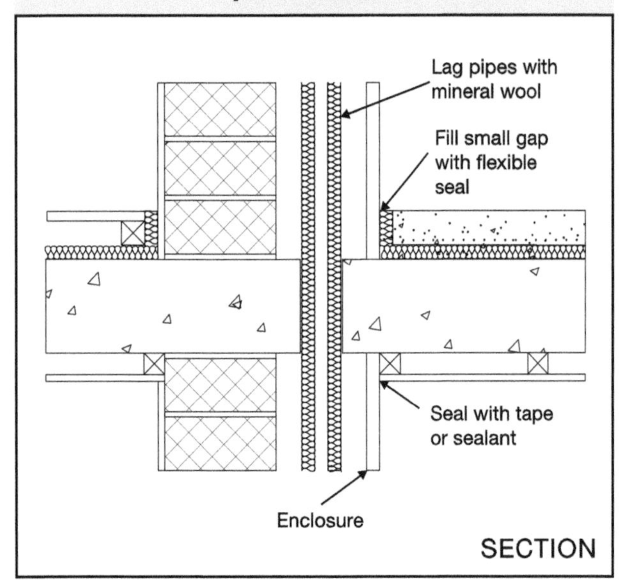

For flats where there are separating walls the following may also apply:

Junctions with a separating wall type 1 – solid masonry

Floor type 2.1C

3.83 A separating floor type 2.1C base (excluding any screed) should pass through a separating wall type 1.

Floor type 2.2B

3.84 A separating floor type 2.2B base (excluding any screed) should not be continuous through a separating wall type 1. See Diagram 3.18.

Diagram 3.18 Floor types 2.2B(a) and 2.2B(b) – wall type 1

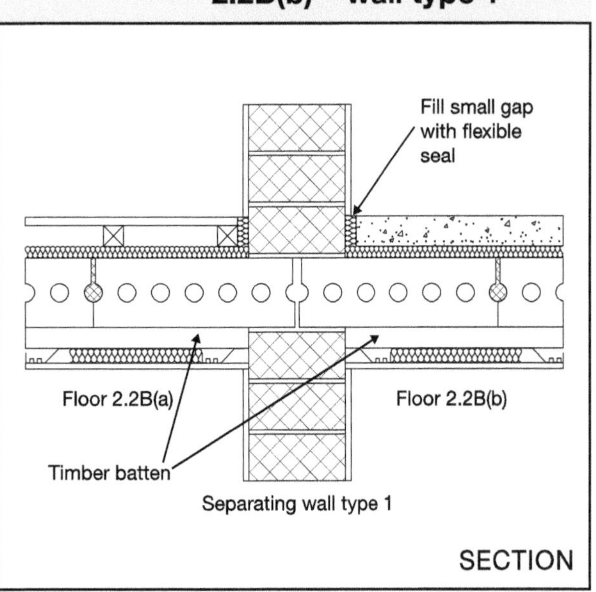

Junctions with a separating wall type 2 – cavity masonry

3.85 The floor base (excluding any screed) should be carried through to the cavity face of the leaf. The cavity should not be bridged.

Floor type 2.2B

3.86 Where floor type 2.2B is used and the planks are parallel to the separating wall the first joint should be a minimum of 300mm from the cavity face of the leaf.

Junctions with a separating wall type 3 – masonry between independent panels

Junctions with separating wall type 3.1 and 3.2 (solid masonry core)

Floor type 2.1C

3.87 A separating floor type 2.1C base (excluding any screed) should pass through separating wall types 3.1 and 3.2. See Diagram 3.19.

Diagram 3.19 Floor type 2.1C – wall types 3.1 and 3.2

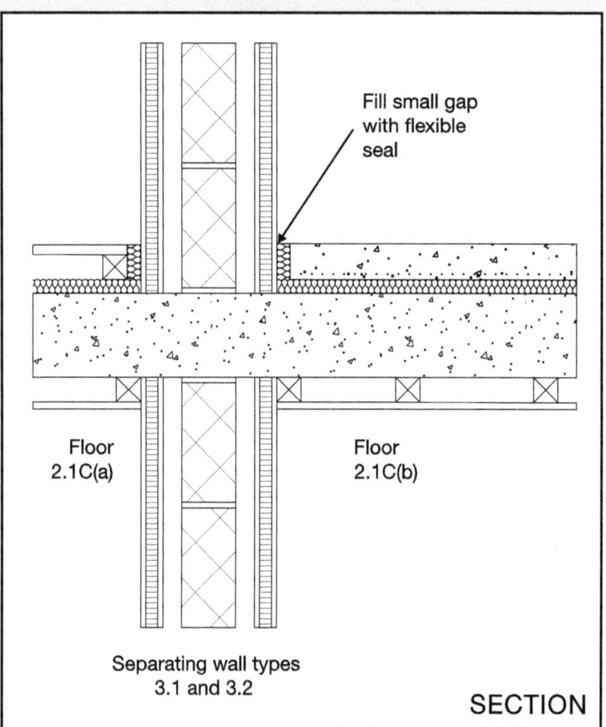

Floor type 2.2B

3.88 A separating floor type 2.2B base (excluding any screed) should not be continuous through a separating wall type 3.

3.89 Where separating wall type 3.2 is used with floor type 2.2B and the planks are parallel to the separating wall the first joint should be a minimum of 300mm from the centreline of the masonry core.

Junctions with separating wall type 3.3 (cavity masonry core)

3.90 The mass per unit area of any leaf that is supporting or adjoining the floor should be at least 120kg/m² excluding finish.

3.91 The floor base (excluding any screed) should be carried through to the cavity face of the leaf of the core. The cavity should not be bridged.

Floor type 2.2B

3.92 Where floor type 2.2B is used and the planks are parallel to the separating wall the first joint should be a minimum of 300mm from the inner face of the adjacent cavity leaf of the masonry core.

Junctions with separating wall type 4 – timber frames with absorbent material

3.93 No guidance available (seek specialist advice).

Floor type 3: timber frame base with ceiling and platform floor

3.94 The resistance to airborne and impact sound depends on the structural floor base and the isolation of the platform floor and the ceiling. The platform floor reduces impact sound at source.

Construction

3.95 The construction consists of a timber frame structural floor base with a deck, platform floor and ceiling treatment A. The platform floor consists of a floating layer and a resilient layer.

3.96 One floor type 3 construction (type 3.1A) is described in this guidance.

3.97 Details of how junctions should be made to limit flanking transmission are also described in this guidance.

Limitations

3.98 Where resistance to airborne sound only is required the full construction should still be used.

3.99 Points to watch

Do

a. Do give special attention to workmanship and detailing at the perimeter and wherever the floor is penetrated, to reduce flanking transmission and to avoid air paths.

b. Do control flanking transmission from walls connected to the separating floor as described in the guidance on junctions.

Platform floor

c. Do use the correct density of resilient layer and ensure it can carry the anticipated load.

d. Do use an expanded or extruded polystyrene strip (or similar resilient material) around the perimeter which is approx. 4mm higher than the upper surface of the floating layer to ensure that during construction a gap is maintained between the wall and the floating layer. This gap may be filled with a flexible sealant.

e. Do lay resilient materials in sheets with joints tightly butted and taped.

Do not

Do not bridge between the floating layer and the base or surrounding walls (e.g. with services or fixings that penetrate the resilient layer).

3.100 Floor type 3.1A *Timber frame base with ceiling treatment A and platform floor (see Diagram 3.20)*

- timber joists with a deck;
- the deck should be of any suitable material with a minimum mass per unit area of 20kg/m²;
- platform floor (including resilient layer) essential;
- ceiling treatment A essential.

Diagram 3.20 **Floor type 3.1A**

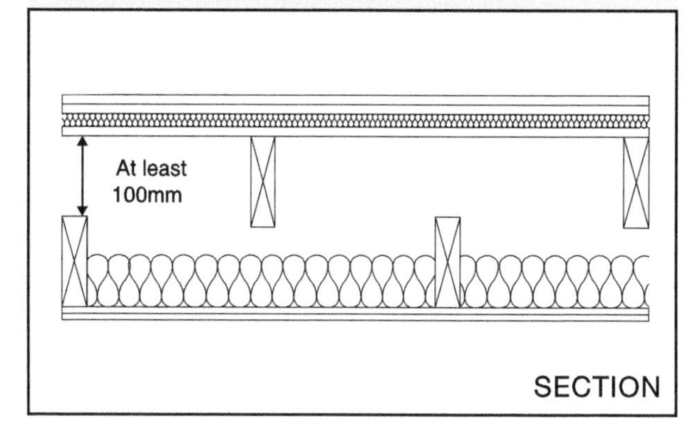

3.101 Platform floor

The floating layer should be:

- a minimum of two layers of board material;
- minimum total mass per unit area 25kg/m^2;
- each layer of minimum thickness 8mm;
- fixed together (e.g. spot bonded with a resilient adhesive or glued/screwed) with joints staggered.

The floating layer should be laid loose on a resilient layer.

Example 1

- 18mm timber or wood based board
- tongued and grooved edges and glued joints
- spot bonded to a substrate of 19mm plasterboard with joints staggered
- minimum total mass per unit area 25kg/m^2

Example 2

- two layers of cement bonded particle board with staggered joints
- total thickness 24mm
- boards glued and screwed together
- minimum total mass per unit area 25kg/m^2

3.102 Resilient layer

The resilient layer specification is:

- mineral wool, minimum thickness 25mm, density 60 to 100kg/m^3;
- the mineral wool may be paper faced on the underside.

Note: The lower figure of density for the resilient layer gives a higher resistance to impact sound but a 'softer' floor. In such cases additional support can be provided around the perimeter of the floor by using a timber batten with a foam strip along the top attached to the wall.

Junction requirements for floor type 3

Junctions with an external cavity wall with masonry inner leaf

3.103 Where the external wall is a cavity wall:

a. the outer leaf of the wall may be of any construction; and

b. the cavity should be stopped with a flexible closer unless the cavity is fully filled with mineral wool or expanded polystyrene beads (seek manufacturer's advice for other suitable materials).

3.104 The masonry inner leaf of a cavity wall should be lined with an independent panel as described for wall type 3.

3.105 The ceiling should be taken through to the masonry. The junction between the ceiling and the independent panel should be sealed with tape or caulked with sealant.

3.106 Use any normal method of connecting floor base to wall but block air paths between floor and wall cavities.

3.107 Where the mass per unit area of the inner leaf is greater than 375kg/m^2 the independent panels are not required.

3.108 See details in Section 2 concerning the use of wall ties in external masonry cavity walls.

Junctions with an external cavity wall with timber frame inner leaf

3.109 Where the external wall is a cavity wall:

a. the outer leaf of the wall may be of any construction; and

b. the cavity should be stopped with a flexible closer.

3.110 The wall finish of the inner leaf of the external wall should be:

a. two layers of plasterboard;

b. each sheet of plasterboard of minimum mass per unit area 10kg/m^2; and

c. all joints should be sealed with tape or caulked with sealant.

3.111 Use any normal method of connecting floor base to wall. Where the joists are at right angles to the wall, spaces between the floor joists should be sealed with full depth timber blocking.

3.112 The junction between the ceiling and wall lining should be sealed with tape or caulked with sealant.

Junctions with an external solid masonry wall

3.113 No guidance available (seek specialist advice).

Junctions with internal framed walls

3.114 Where the joists are at right angles to the wall, spaces between the floor joists should be sealed with full depth timber blocking.

3.115 The junction between the ceiling and the internal framed wall should be sealed with tape or caulked with sealant.

Junctions with internal masonry walls

3.116 No guidance available (seek specialist advice).

Junctions with floor penetrations (excluding gas pipes)

3.117 Pipes and ducts that penetrate a floor separating habitable rooms in different flats should be enclosed for their full height in each flat. See Diagram 3.21.

3.118 The enclosure should be constructed of material having a mass per unit area of at least 15kg/m^2. Either line the enclosure, or wrap the duct or pipe within the enclosure, with 25mm unfaced mineral wool.

3.119 Leave a small gap (approx. 5mm) between enclosure and floating layer and seal with sealant or neoprene. The enclosure may go down to the floor base, but ensure that the enclosure is isolated from the floating layer.

3.120 Penetrations through a separating floor by ducts and pipes should have fire protection to satisfy Building Regulation Part B – Fire safety. Fire stopping should be flexible and also prevent rigid contact between the pipe and floor.

Note: There are requirements for ventilation of ducts at each floor where they contain gas pipes. Gas pipes may be contained in a separate ventilated duct or they can remain unenclosed. Where a gas service is installed, it shall comply with relevant codes and standards to ensure safe and satisfactory operation. See The Gas Safety (Installation and Use) Regulations 1998, SI 1998/2451.

Diagram 3.21 **Floor type 3 – floor penetrations**

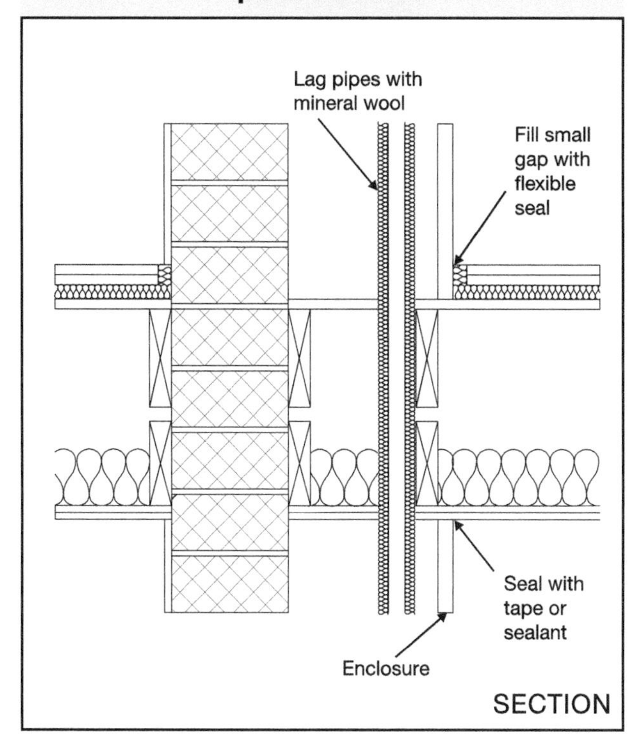

For flats where there are separating walls the following may also apply:

Junctions with a separating wall type 1 – solid masonry

3.121 If floor joists are to be supported on the separating wall then they should be supported on hangers and should not be built in. See Diagram 3.22.

3.122 The junction between the ceiling and wall should be sealed with tape or caulked with sealant.

Diagram 3.22 **Floor type 3 – wall type 1**

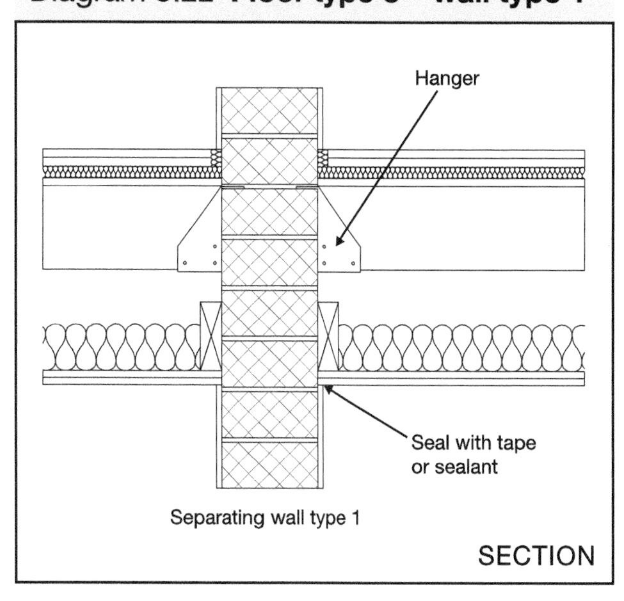

Junctions with a separating wall type 2 – cavity masonry

3.123 If floor joists are to be supported on the separating wall then they should be supported on hangers and should not be built in. See Diagram 3.23.

3.124 The adjacent leaf of a cavity separating wall should be lined with an independent panel as described in wall type 3.

3.125 The ceiling should be taken through to the masonry. The junction between the ceiling and the independent panel should be sealed with tape or caulked with sealant.

3.126 Where the mass per unit area of the adjacent leaf is greater than 375kg/m^2 the independent panels are not required.

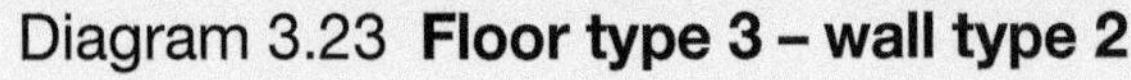

Diagram 3.23 **Floor type 3 – wall type 2**

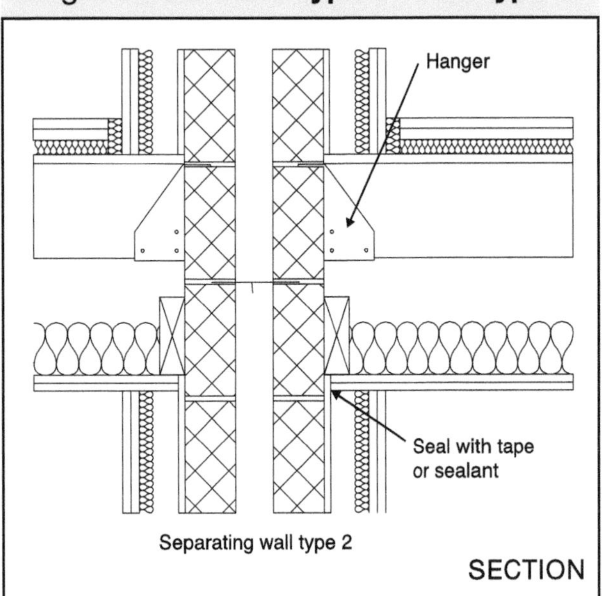

Junctions with a separating wall type 3 – masonry between independent panels

3.127 If floor joists are to be supported on the separating wall then they should be supported on hangers and should not be built in.

3.128 The ceiling should be taken through to the masonry. The junction between the ceiling and the independent panel should be sealed with tape or caulked with sealant.

Junctions with a separating wall type 4 – timber frames with absorbent material

3.129 Where the joists are at right angles to the wall, spaces between the floor joists should be sealed with full depth timber blocking.

3.130 The junction of the ceiling and wall lining should be sealed with tape or caulked with sealant.

Section 4: Dwelling-houses and flats formed by material change of use

Introduction

4.1 This Section gives guidance on dwelling-houses and flats formed by material change of use. For rooms for residential purposes formed by material change of use see Section 6.

4.2 It may be that an existing wall, floor or stair in a building that is to undergo a material change of use will achieve the performance standards set out in Section 0: Performance – Table 1a without the need for remedial work. This would be the case if the construction was generally similar (including flanking constructions) to one of the constructions in Sections 2 and 3 (e.g. concerning the mass requirement, the structure under consideration should be within 15% of the mass per unit area of a construction listed in the relevant section).

4.3 In other circumstances it may be possible to use the guidance in Section 2 or 3 (including flanking constructions) to determine the appropriate remedial treatment which will result in the construction achieving the performance standards in Section 0: Performance – Table 1a.

4.4 For situations where it is uncertain whether the existing construction achieves the performance standards set out in Section 0: Performance – Table 1a, this section describes one wall treatment, two floor treatments and one stair treatment as shown in Diagram 4.1. These constructions can be used to increase the sound insulation.

4.5 The guidance in this section is not exhaustive and other designs, materials or products may be used to achieve the performance standards set out in Section 0: Performance – Table 1a. Advice should be sought from the manufacturer or other appropriate source.

4.6 Wall treatment 1 *Independent panel(s) with absorbent material*

The resistance to airborne sound depends on the form of existing construction, the mass of the independent panel(s), the isolation of the panel(s) and the absorbent material.

4.7 Floor treatment 1 *Independent ceiling with absorbent material*

The resistance to airborne and impact sound depends on the combined mass of the existing floor and the independent ceiling, the absorbent material, the isolation of the independent ceiling and the airtightness of the whole construction.

4.8 Floor treatment 2 *Platform floor with absorbent material*

The resistance to airborne and impact sound depends on the total mass of the floor, the effectiveness of the resilient layer and the absorbent material.

4.9 Stair treatment 1 *Stair covering and independent ceiling with absorbent material*

To be used where a timber stair performs a separating function. The resistance to airborne sound depends mainly on the mass of the stair, the mass and isolation of any independent ceiling and the airtightness of any cupboard or enclosure under the stairs. The stair covering reduces impact sound at source.

4.10 In all cases it may be necessary to control flanking transmission in order to achieve the performance standards set out in Section 0: Performance – Table 1a. See Section 4: Junction requirements for material change of use.

4.11 Special attention needs to be given to situations where flanking walls or floors are continuous across separating walls or floors as a result of the conversion work. In such instances additional treatments may be required to control flanking transmission along these continuous elements. Specialist advice may be needed.

4.12 Significant differences may frequently occur between the construction and layout of each converted unit in a development. Building control bodies should have regard to the guidance in Section 1 when deciding on the application of pre-completion testing to material change of use.

4.13 For some historic buildings undergoing a material change of use, it may not be practical to improve the sound insulation to the performance standards set out in Section 0: Performance – Table 1a. In such cases refer to Section 0: Performance, paragraph 0.7.

4.14 Wall and floor treatments will impose additional loads on the existing structure. The structure should be assessed to ensure that the additional loading can be carried safely, with appropriate strengthening applied where necessary.

4.15 Floor or wall penetrations, such as ducts or pipes, passing through separating elements in conversions can reduce the level of sound insulation. Guidance on the treatment of floor penetrations is given below.

Work to existing construction

4.16 Before a floor treatment is applied appropriate remedial work to the existing construction should be undertaken as described in paragraphs 4.17 and 4.18.

4.17 If the existing floor is timber then gaps in floor boarding should be sealed by overlaying with hardboard or filled with sealant.

a. Where floor boards are to be replaced, boarding should have a minimum thickness of 12mm, and mineral wool (minimum thickness 100mm, minimum density 10kg/m^3) should be laid between the joists in the floor cavity.

b. If the existing floor is concrete and the mass per unit area of the concrete floor is less than 300kg/m^2, or is unknown, then the mass of the floor should be increased to at least 300kg/m^2. Any air gaps through a concrete floor should be sealed. A regulating screed may also be required.

c. If there is an existing lath and plaster ceiling it should be retained as long as it satisfies Building Regulation Part B – Fire safety.

d. Where the existing ceiling is not lath and plaster it should be upgraded as necessary to provide at least two layers of plasterboard with joints staggered, total mass per unit area 20kg/m^2.

4.18 Extensive remedial work to reduce flanking transmission may also be necessary to achieve the performance standards set out in Section 0: Performance – Table 1a. This may involve wall linings, see Section 4: Junction requirements for material change of use, paragraphs 4.43 and 4.44.

Corridor walls and doors

4.19 The separating walls described in this section should be used between dwelling-houses, or flats formed by material change of use, and corridors in order to control flanking transmission and to provide the required sound insulation. However, it is likely that the sound insulation will be reduced by the presence of a door.

4.20 Ensure that any door has good perimeter sealing (including the threshold where practical) and a minimum mass per unit area of 25kg/m^2 or a minimum sound reduction index of 29dB R_w (measured according to BS EN ISO 140-3:1995 and rated according to BS EN ISO 717-1:1997). The door should also satisfy the Requirements of Building Regulation Part B – Fire safety.

4.21 Noisy parts of the building should preferably have a lobby, double door or high performance doorset to contain the noise. Where this is not possible, nearby flats should have similar protection. However, there should be a sufficient number of them that are suitable for disabled access, see Building Regulations Part M – Access and facilities for disabled people.

Diagram 4.1 **Treatments for material change of use**

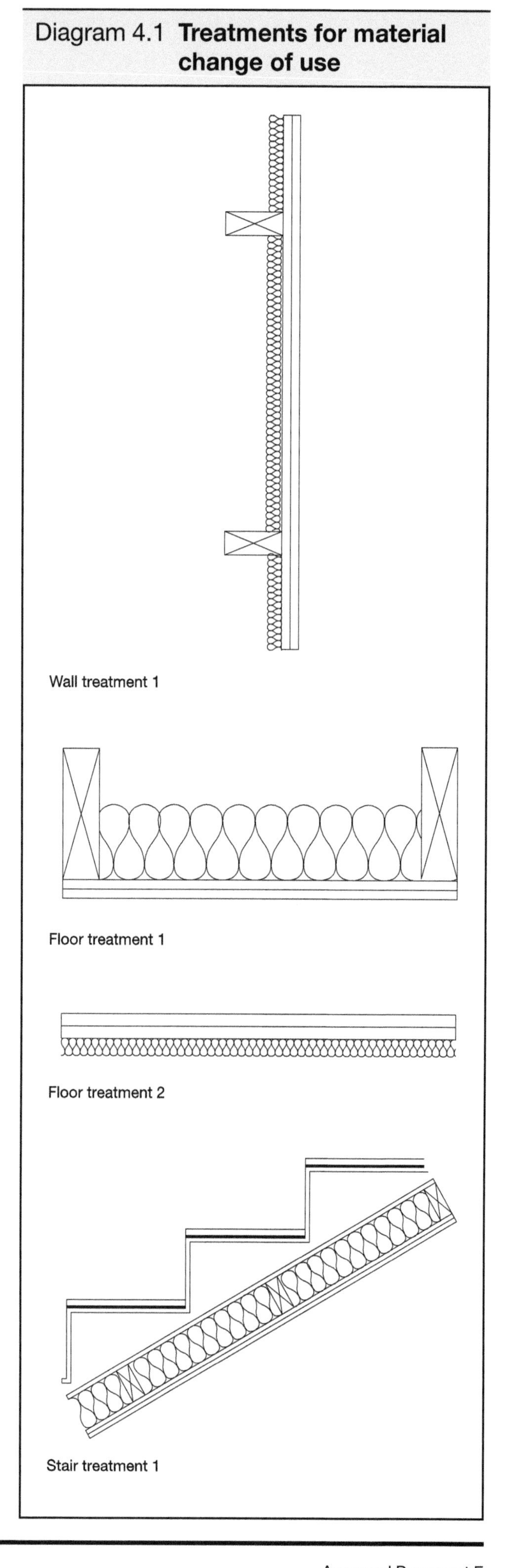

Wall treatment 1: independent panel(s) with absorbent material

4.22 The resistance to airborne sound depends on the form of existing construction, the mass of independent panel(s), the isolation of the panel(s) and the absorbent material.

Construction

4.23 The independent panel may be used on one side of the existing wall only where the existing wall is masonry, and has a thickness of at least 100mm and is plastered on both faces. With other types of existing wall the independent panels should be built on both sides.

4.24 Independent panel(s) with absorbent material (see Diagram 4.2)

- minimum mass per unit area of panel (excluding any supporting framework) 20kg/m^2;
- each panel should consist of at least two layers of plasterboard with staggered joints;
- if the panels are free-standing they should be at least 35mm from masonry core;
- if the panels are supported on a frame there should be a gap of at least 10mm between the frame and the face of the existing wall;
- mineral wool, minimum density 10kg/m^3 and minimum thickness 35mm, in the cavity between the panel and the existing wall.

4.25 Points to watch:

Do

a. Do ensure that the independent panel and its supporting frame are not in contact with the existing wall.

b. Do seal the perimeter of the independent panel with tape or sealant.

Do not

Do not tightly compress the absorbent material as this may bridge the cavity.

Diagram 4.2 **Wall treatment 1**

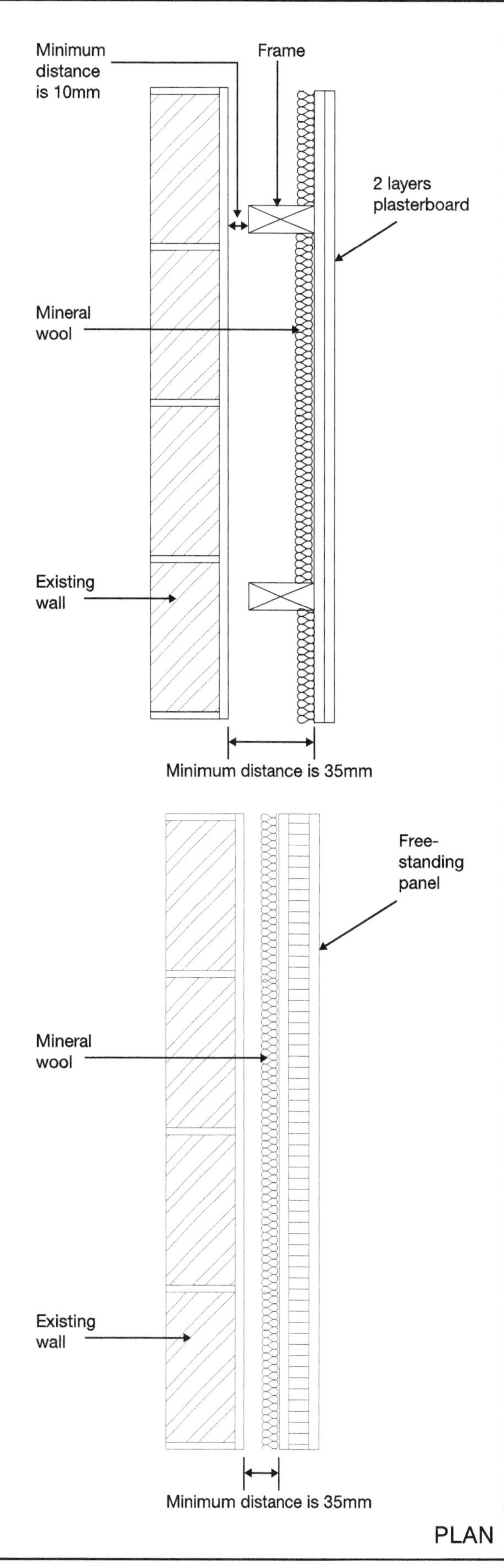

Floor treatment 1: independent ceiling with absorbent material

4.26 The resistance to airborne and impact sound depends on the combined mass of the existing floor and the independent ceiling, the absorbent material, the isolation of the independent ceiling and the airtightness of the whole construction.

4.27 Independent ceiling with absorbent material (see Diagram 4.3)

- at least 2 layers of plasterboard with staggered joints, minimum total mass per unit area 20kg/m^2;
- an absorbent layer of mineral wool laid on the ceiling, minimum thickness 100mm, minimum density 10kg/m^3.

The ceiling should be supported by one of the following methods:

- independent joists fixed only to the surrounding walls. A clearance of at least 25mm should be left between the top of the independent ceiling joists and the underside of the existing floor construction; or
- independent joists fixed to the surrounding walls with additional support provided by resilient hangers attached directly to the existing floor base.

Note: This construction involves a separation of at least 125mm between the upper surface of the independent ceiling and the underside of the existing floor construction. However, structural considerations determining the size of ceiling joists will often result in greater separation. Care should be taken at the design stage to ensure that adequate ceiling height is available in all rooms to be treated.

4.28 Where a window head is near to the existing ceiling, the new independent ceiling may be raised to form a pelmet recess. See Diagram 4.4.

4.29 For the junction detail between floor treatment 1 and wall treatment 1, see Diagram 4.5.

4.30 Points to watch:

Do

a. Do remember to apply appropriate remedial work to the existing construction.

b. Do seal the perimeter of the independent ceiling with tape or sealant.

Do not

a. Do not create a rigid or direct connection between the independent ceiling and the floor base.

b. Do not tightly compress the absorbent material as this may bridge the cavity.

Diagram 4.3 **Floor treatment 1**

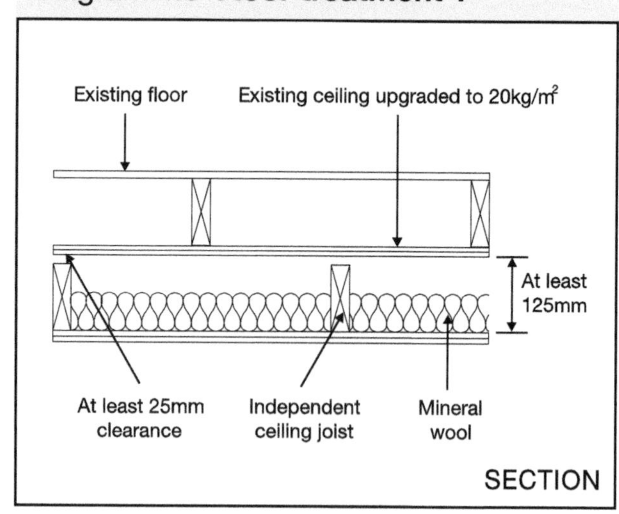

Diagram 4.4 **Floor treatment 1 – high window head detail**

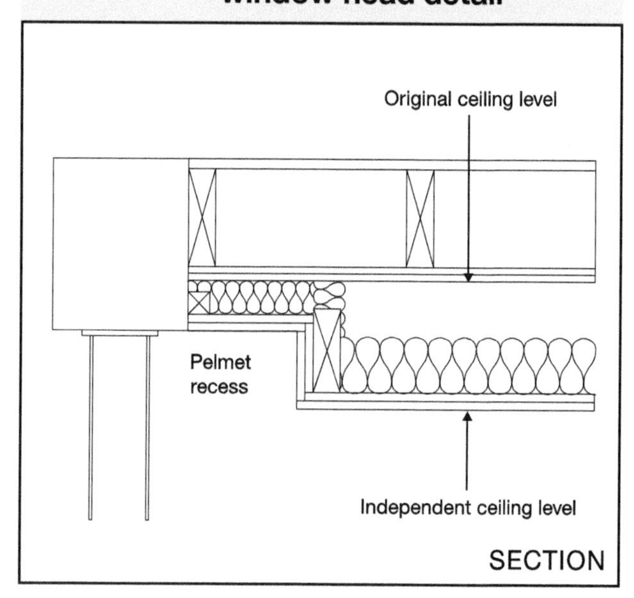

Diagram 4.5 **Floor treatment 1 – wall treatment 1**

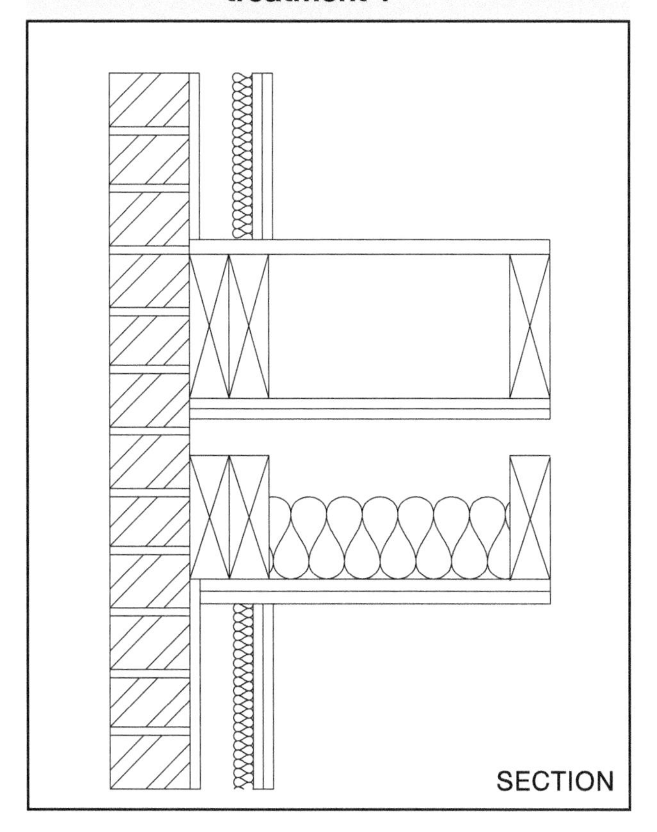

Floor treatment 2: platform floor with absorbent material

4.31 The resistance to airborne and impact sound depends on the total mass of the floor, the effectiveness of the resilient layer and the absorbent material.

4.32 Platform floor with absorbent material (see Diagram 4.6)

Where this treatment is used to improve an existing timber floor, a layer of mineral wool (minimum thickness 100mm, minimum density 10kg/m³) should be laid between the joists in the floor cavity.

The floating layer should be:

- a minimum of two layers of board material;
- minimum total mass per unit area 25kg/m²;
- each layer of minimum thickness 8mm;
- fixed together (e.g. spot bonded or glued/ screwed) with joints staggered.

The floating layer should be laid loose on a resilient layer. The resilient layer specification is:

- mineral wool, minimum thickness 25mm, density 60 to 100kg/m³;
- the mineral wool may be paper faced on the underside.

Note: The lower figure of density for the resilient layer gives the best insulation but a 'softer' floor. In such cases additional support can be provided around the perimeter of the floor by using a timber batten with a foam strip along the top attached to the wall.

4.33 For the junction detail between floor treatment 2 and wall treatment 1, see Diagram 4.7.

4.34 Points to watch:

Do

a. Do remember to apply appropriate remedial work to the existing construction.

b. Do use the correct density of resilient layer and ensure it can carry the anticipated load.

c. Do allow for movement of materials e.g. expansion of chipboard after laying (to maintain isolation).

d. Do carry the resilient layer up at all room edges to isolate the floating layer from the wall surface.

e. Do leave a small gap (approx. 5mm) between skirting and floating layer and fill with a flexible sealant.

f. Do lay resilient materials in sheets with joints tightly butted and taped.

g. Do seal the perimeter of any new ceiling with tape or sealant.

Do not

Do not bridge between the floating layer and the base or surrounding walls (e.g. with services or fixings that penetrate the resilient layer).

Diagram 4.6 **Floor treatment 2**

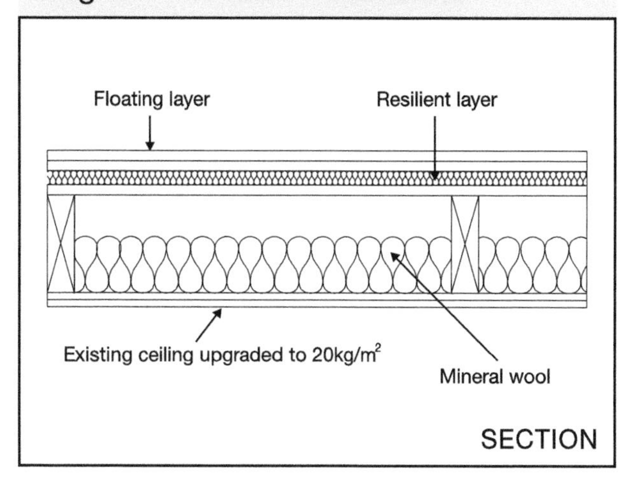

Diagram 4.7 **Floor treatment 2 – wall treatment 1**

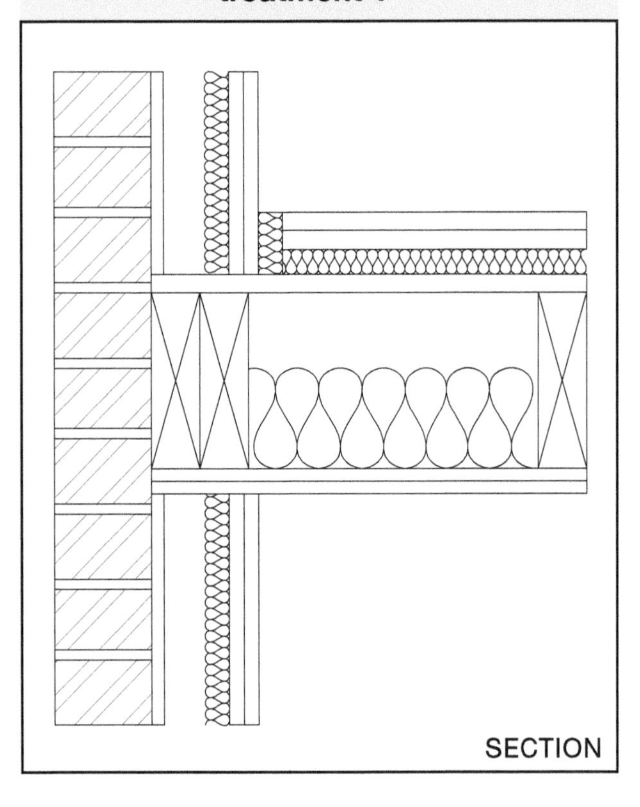

Where there is no cupboard under the stair construct an independent ceiling below the stair (see Floor treatment 1).

4.38 For fire protection where a staircase performs a separating function refer to Building Regulation Part B – Fire safety.

Diagram 4.8 **Stair treatment**

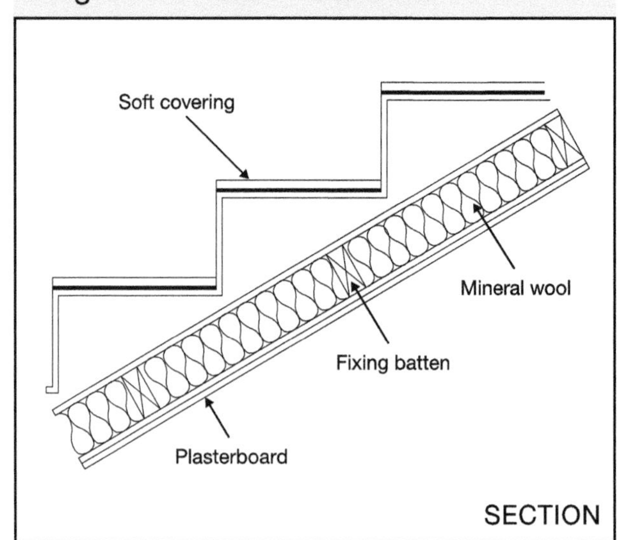

Stair treatment: stair covering and independent ceiling with absorbent material

4.35 Stairs are subject to the same sound insulation requirements as floors where they perform a separating function.

4.36 The resistance to airborne sound depends mainly on the mass of the stair, the mass and isolation of any independent ceiling and the airtightness of any cupboard or enclosure under the stairs. The stair covering reduces impact sound at source.

4.37 Stair covering and independent ceiling with absorbent material

Lay soft covering of at least 6mm thickness over the stair treads. Ensure it is securely fixed (e.g. glued) so it does not become a safety hazard.

If there is a cupboard under all, or part, of the stair:

a. line the underside of the stair within the cupboard with plasterboard of minimum mass per unit area 10kg/m^2 and an absorbent layer of mineral wool (minimum density 10kg/m^3), within the space above the lining; and

b. build cupboard walls from two layers of plasterboard (or equivalent), each sheet of minimum mass per unit area 10kg/m^2; and

c. use a small, heavy, well fitted door for the cupboard.

Junction requirements for material change of use

Junctions with abutting construction

4.39 For floating floors, carry the resilient layer up at all room edges to isolate the floating layer from the wall surface.

4.40 For floating floors, leave a small gap (approx. 5mm) between the skirting and floating layer and fill with a flexible sealant.

4.41 The perimeter of any new ceiling should be sealed with tape or caulked with sealant.

4.42 Relevant junction details are shown in Diagrams 4.5 and 4.7.

Junctions with external or load-bearing walls

4.43 Where there is significant flanking transmission along adjoining walls then improved sound insulation can be achieved by lining all adjoining masonry walls with either

a. an independent layer of plasterboard; or

b. a laminate of plasterboard and mineral wool. For other drylining laminates, seek advice from the manufacturer.

4.44 Where the adjoining masonry wall has a mass per unit area greater than 375kg/m^2 then such lining may not be necessary, as it may not give a significant improvement.

Note: Specialist advice may be needed on the diagnosis and control of flanking transmission.

Junctions with floor penetrations

4.45 Piped services (excluding gas pipes) and ducts which pass through separating floors in conversions should be surrounded with sound absorbent material for their full height and enclosed in a duct above and below the floor.

Do

a. Do seal the joint between casings and ceiling with tape or sealant.

b. Do leave a nominal gap (approx. 5mm) between the casing and any floating layer and fill with sealant.

Construction

4.46 Pipes and ducts that penetrate a floor separating habitable rooms in different flats should be enclosed for their full height in each flat.

4.47 The enclosure should be constructed of material having a mass per unit area of at least 15kg/m^2.

4.48 Either line the enclosure, or wrap the duct or pipe within the enclosure, with 25mm unfaced mineral wool.

4.49 The enclosure may go down to the floor base if floor treatment 2 is used but ensure isolation from the floating layer.

4.50 Penetrations through a separating floor by ducts and pipes should have fire protection to satisfy Building Regulation Part B – Fire safety. Fire stopping should be flexible and also prevent rigid contact between the pipe and floor.

Note: There are requirements for ventilation of ducts at each floor where they contain gas pipes. Gas pipes may be contained in a separate ventilated duct or they can remain unducted. Where a gas service is installed, it shall comply with relevant codes and standards to ensure safe and satisfactory operation. See The Gas Safety (Installation and Use) Regulations 1998, SI 1998/2451.

Diagram 4.9 **Floor penetrations**

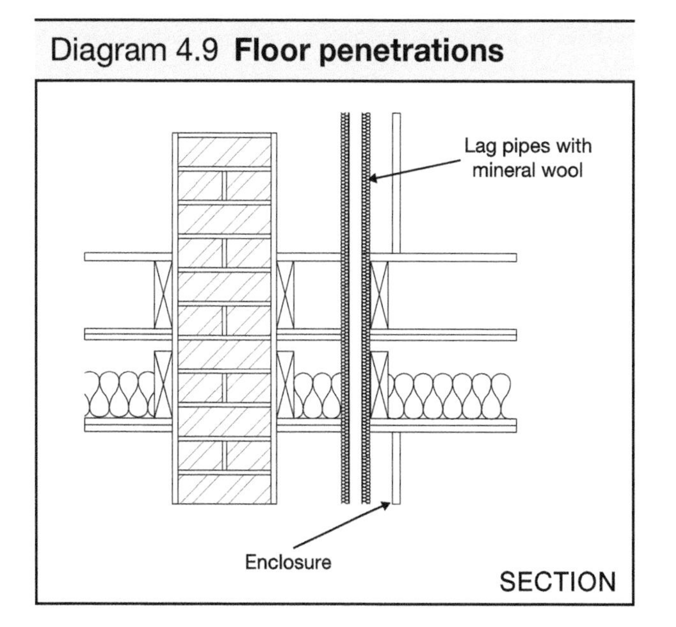

Section 5: Internal walls and floors for new buildings

Introduction

5.1 This Section gives examples of internal wall and floor constructions that meet the laboratory sound insulation values set out in Section 0: Performance – Table 2.

5.2 These constructions have been designed to give insulation against airborne sound. For internal floors, insulation against impact sound could be improved by adding a soft covering (e.g. carpet).

5.3 They are grouped in four main types as shown below.

5.4 Internal wall type A or B: *Timber or metal frame*

The resistance to airborne sound depends on the mass per unit area of the leaves, the cavity width, frame material and the absorption in the cavity between the leaves.

5.5 Internal wall type C or D: *Concrete or aircrete block*

The resistance to airborne sound depends mainly on the mass per unit area of the wall.

5.6 Internal floor type A or B: *Concrete planks or concrete beams with infilling blocks*

The resistance to airborne sound depends on the mass per unit area of the concrete base or concrete beams and infilling blocks. A soft covering will reduce impact sound at source.

5.7 Internal floor type C: *Timber or metal joist*

The resistance to airborne sound depends on the structural floor base, the ceiling and the absorbent material. A soft covering will reduce impact sound at source.

5.8 For both internal walls and internal floors the constructions are ranked, as far as possible, with constructions giving better sound insulation given first.

Doors

5.9 Lightweight doors with poor perimeter sealing provide a lower standard of sound insulation than walls. This will reduce the effective sound insulation of the internal wall. Ways of improving sound insulation include ensuring that there is good perimeter sealing or by using a doorset.

5.10 See Building Regulation Part F – Ventilation and Part J – Combustion appliances and fuel storage systems.

Layout

5.11 If the stair is not enclosed, then the potential sound insulation of the internal floor will not be achieved; nevertheless, the internal floor should still satisfy Requirement E2.

5.12 It is good practice to consider the layout of rooms at the design stage to avoid placing noise sensitive rooms next to rooms in which noise is generated. Guidance on layout is provided in BS 8233:1999 Sound Insulation and Noise Reduction for Buildings. Code of Practice.

Junction requirements for internal walls

5.13 Section 3: Separating Floors contains important guidance on junctions of separating floors with internal walls.

5.14 Fill all gaps around internal walls to avoid air paths between rooms.

Junction requirements for internal floors

5.15 Section 2: Separating Walls contains important guidance on junctions of separating walls with internal floors.

5.16 Fill all gaps around internal floors to avoid air paths between rooms.

5.17 Internal wall type A: *Timber or metal frames with plasterboard linings on each side of frame (see Diagram 5.1)*

- each lining to be two or more layers of plasterboard, each sheet of minimum mass per unit area 10kg/m²;
- linings fixed to timber frame with a minimum distance between linings of 75mm, or metal frame with a minimum distance between linings of 45mm;
- all joints well sealed.

Diagram 5.1 **Internal wall type A**

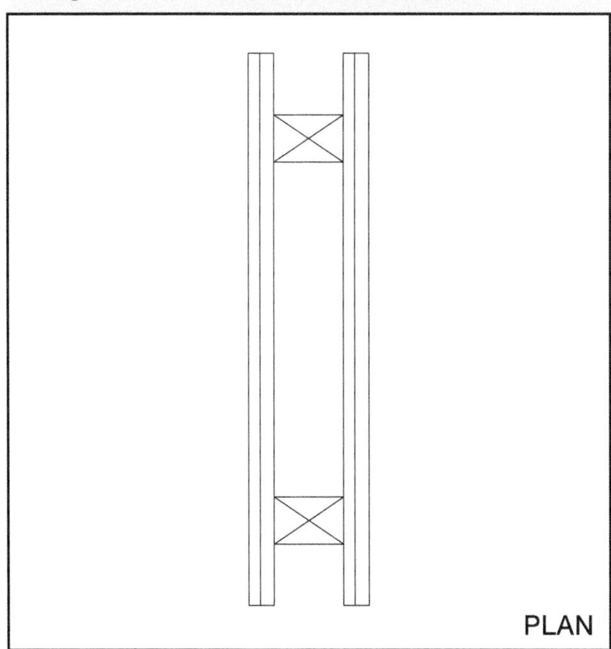

5.18 Internal wall type B: *Timber or metal frames with plasterboard linings on each side of frame and absorbent material (see Diagram 5.2)*

- single layer of plasterboard of minimum mass per unit area 10kg/m²;
- linings fixed to timber frame with a minimum distance between linings of 75mm, or metal frame with a minimum distance between linings of 45mm;
- an absorbent layer of unfaced mineral wool batts or quilt (minimum thickness 25mm, minimum density 10kg/m³) which may be wire reinforced, suspended in the cavity;
- all joints well sealed.

Diagram 5.2 **Internal wall type B**

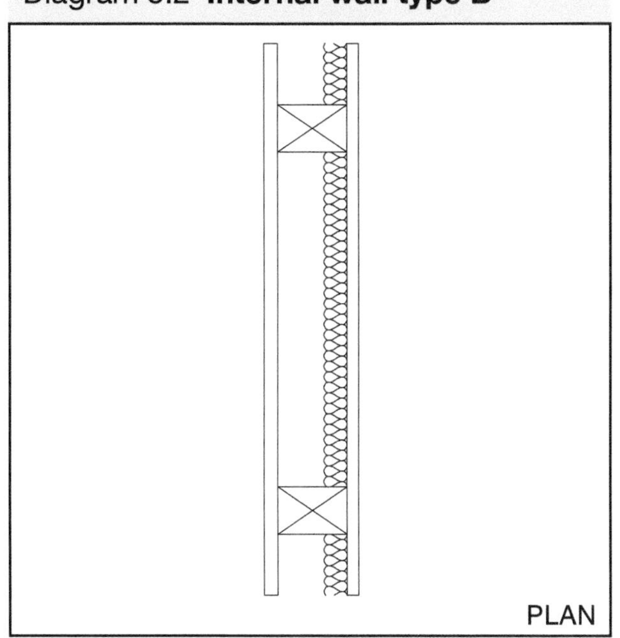

5.19 Internal wall type C: *Concrete block wall, plaster or plasterboard finish on both sides (see Diagram 5.3)*

- minimum mass per unit area, excluding finish 120kg/m²;
- all joints well sealed;
- plaster or plasterboard finish on both sides.

Diagram 5.3 **Internal wall type C**

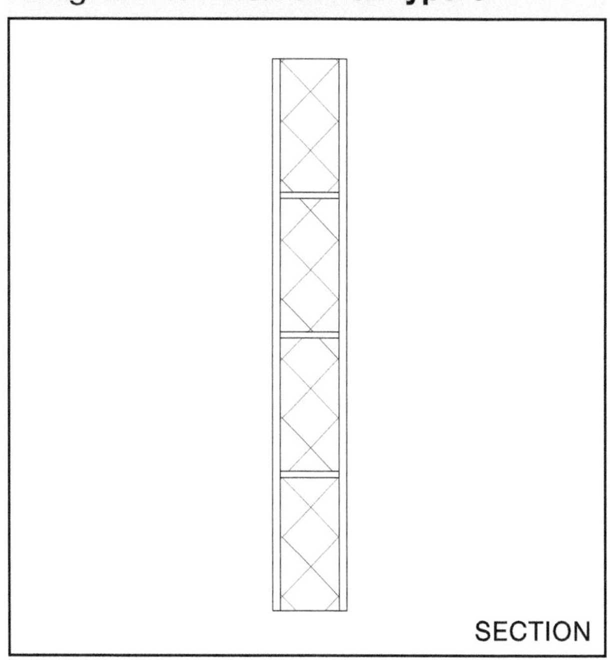

5.20 Internal wall type D: *Aircrete block wall, plaster or plasterboard finish on both sides (see Diagram 5.4)*

- for plaster finish, minimum mass per unit area, including finish 90kg/m²;
- for plasterboard finish, minimum mass per unit area, including finish 75kg/m²;
- all joints well sealed;
- internal wall type D should only be used with the separating walls described in this Approved Document where there is no minimum mass requirement on the internal masonry walls. See guidance in Section 2;
- internal wall type D should not be used as a load-bearing wall connected to a separating floor, or be rigidly connected to the separating floors described in this Approved Document. See guidance in Section 3.

Diagram 5.4 **Internal wall type D**

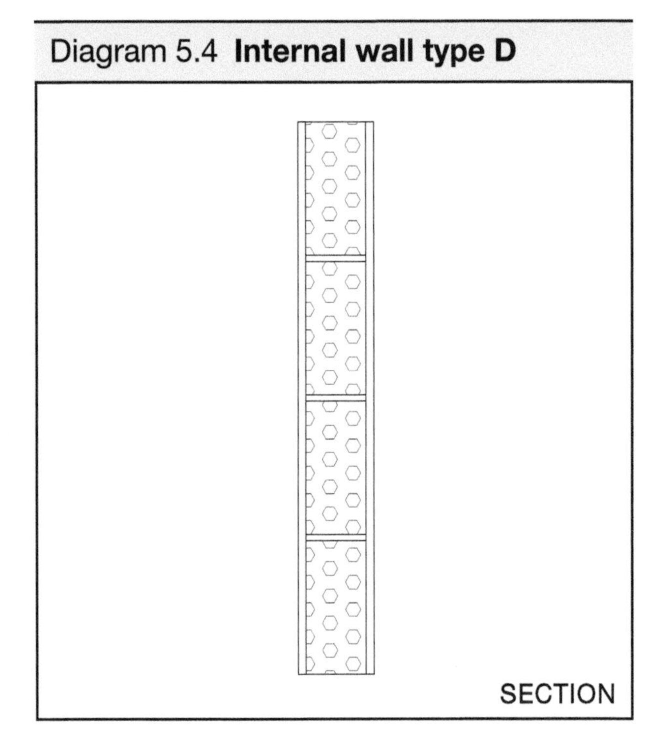

5.21 Internal floor type A: *Concrete planks (see Diagram 5.5)*

- minimum mass per unit area 180kg/m²;
- regulating screed optional;
- ceiling finish optional.

Note: Insulation against impact sounds can be improved by adding a soft covering (e.g. carpet).

Diagram 5.5 **Internal floor type A**

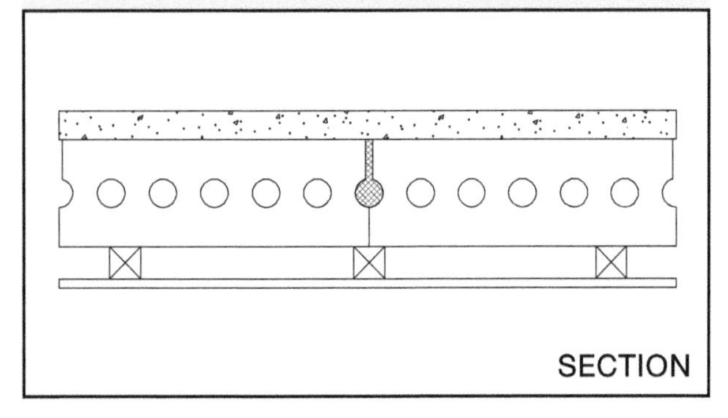

5.22 Internal floor type B: *Concrete beams with infilling blocks, bonded screed and ceiling (see Diagram 5.6)*

- minimum mass per unit area of beams and blocks 220kg/m²;
- bonded screed required. Sand cement screeds should have a minimum thickness of 40mm. For proprietary bonded screed products, seek manufacturer's advice on the appropriate thickness;
- ceiling finish required. Use ceiling treatment C or better from Section 3.

Note: Insulation against impact sounds can be improved by adding a soft covering (e.g. carpet).

Diagram 5.6 **Internal floor type B**

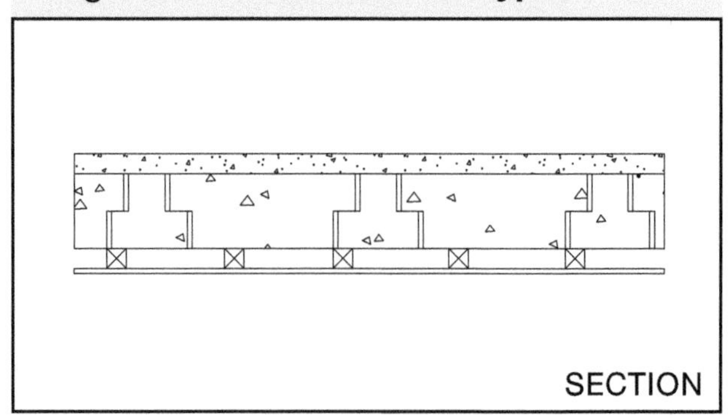

5.23 Internal floor type C: *Timber or metal joist, with wood-based board and plasterboard ceiling, and absorbent material (see Diagram 5.7)*

- floor surface of timber- or wood-based board, minimum mass per unit area 15kg/m²;
- ceiling treatment of single layer of plasterboard, minimum mass per unit area 10kg/m², fixed using any normal fixing method;
- an absorbent layer of mineral wool (minimum thickness 100mm, minimum density 10kg/m³) laid in the cavity.

Note: Insulation against impact sounds can be improved by adding a soft covering (e.g. carpet).

Diagram 5.7 **Internal floor type C**

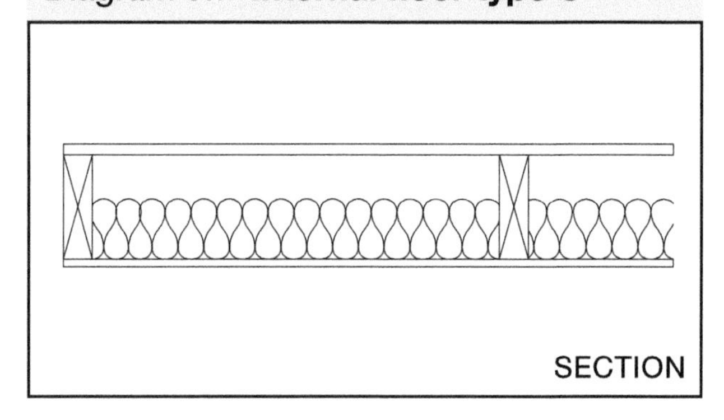

Note: Electrical cables give off heat when in use and special precautions may be required when they are covered by thermally insulating materials. See BRE BR 262, Thermal Insulation: avoiding risks, Section 2.4.

Section 6: Rooms for residential purposes

Introduction

6.1 Rooms for residential purposes are defined in Regulation 2 of the Building Regulations 2010. This definition is reproduced after the Requirements in this Approved Document.

6.2 This Section gives examples of wall and floor types, which, if built correctly, should meet the performance standards set out in Section 0: Performance – Table 1b.

6.3 The guidance in this section is not exhaustive and other designs, materials or products may be used to achieve the performance standards set out in Section 0: Performance – Table 1b. Advice should be sought from the manufacturer or other appropriate source.

Separating walls in new buildings containing rooms for residential purposes

6.4 Of the separating walls described in Section 2 the following types are most suitable for use in new buildings containing rooms for residential purposes:

Wall type 1. Solid masonry

- Wall type 1.1, Dense aggregate concrete block, plaster on both room faces;
- Wall type 1.2, Dense aggregate concrete in situ, plaster on both room faces;
- Wall type 1.3, Brick, plaster on both room faces.

Note: Plasterboard may be used as an alternative wall finish, provided a sheet of minimum mass per unit area 10kg/m^2 is used on each room face.

Wall type 3. Masonry between independent panels

- Wall type 3.1, Solid masonry core (dense aggregate concrete block), independent panels on both room faces.
- Wall type 3.2, Solid masonry core (lightweight concrete block), independent panels on both room faces.

Note: Wall types 2 and 4 can be used provided that care is taken to maintain isolation between the leaves. Specialist advice may be needed.

Corridor walls and doors

6.5 Separating walls described in 6.4 should be used between rooms for residential purposes and corridors in order to control flanking transmission and to provide the required sound insulation between the dwelling and the corridor. However, it is likely that the sound insulation will be reduced by the presence of a door.

6.6 Ensure any door has good perimeter sealing (including the threshold where practical) and a minimum mass per unit area of 25kg/m^2. Alternatively, use a doorset with a minimum sound reduction index of 29dB R_w (measured in the laboratory according to BS EN ISO 140-3:1995 and rated according to BS EN ISO 717-1:1997). The door should also satisfy the Requirements of Building Regulation Part B – Fire safety.

6.7 Noisy parts of the building (e.g. function rooms or bars) should preferably have a lobby, double door or high performance doorset to contain the noise. Where this is not possible, nearby rooms for residential purposes should have similar protection. However, do ensure that there are doors that are suitable fordisabled access, see Building Regulations Part M – Access and facilities for disabled people.

Separating floors in new buildings containing rooms for residential purposes

6.8 Of the separating floors described in Section 3 the following types are most suitable for use in new buildings containing rooms for residential purposes:

Floor type 1. Concrete base with soft covering

- Floor type 1.1C Solid concrete slab (cast in situ, with or without permanent shuttering), soft floor covering, ceiling treatment C.
- Floor type 1.2B Concrete planks (solid or hollow), soft floor covering, ceiling treatment B.

Note: Floor types 2 and 3 can be used provided that floating floors and ceilings are not continuous between rooms for residential purposes. Specialist advice may be needed.

Rooms for residential purposes resulting from a material change of use

6.9 It may be that an existing wall, floor or stair in a building that is to undergo a material change of use will achieve the performance standards set out in Section 0: Performance – Table 1b without the need for remedial work. This would be the case if the construction was similar (including flanking constructions) to one of the constructions in paragraphs 6.4 and 6.8 (e.g. for solid walls and floors the mass requirement should be within 15% of the mass per unit area of a construction listed in the relevant section).

6.10 For situations where it cannot be shown that the existing construction will achieve the performance standards set out in Section 0: Performance – Table 1b, Section 4 describes wall, floor and stair treatments to improve the level of sound insulation in dwellings formed by material change of use. These treatments may be used in buildings containing rooms for residential purposes. Specialist advice may be needed.

Junction details

6.11 In order for the construction to be fully effective, care should be taken to detail correctly the junctions between the separating wall and other elements, such as floors, roofs, external walls and internal walls.

6.12 In the case of new buildings containing rooms for residential purposes, refer to the guidance in Sections 2 and 3 which describes the junction and flanking details for each of the new build separating wall and floor types.

6.13 When rooms for residential purposes are formed by material change of use, refer to the notes and diagrams in Section 4 that describe the junction and flanking details for the wall and floor treatments.

6.14 In the case of the junction between a solid masonry separating wall type 1 and the ceiling void and roof space, the solid wall need not be continuous to the underside of the structural floor or roof provided that:

a. there is a ceiling consisting of two or more layers of plasterboard, of minimum total mass per unit area 20kg/m^2;

b. there is a layer of mineral wool (minimum thickness 200mm, minimum density 10kg/m^3) in the roof void;

c. the ceiling is not perforated.

The ceiling joists and plasterboard sheets should not be continuous between rooms for residential purposes. See Diagram 6.1.

6.15 This ceiling void and roof space detail can only be used where the Requirements of Building Regulations Part B – Fire safety can also be satisfied. The Requirements of Building Regulations Part L – Conservation of fuel and power should also be satisfied.

Diagram 6.1 **Ceiling void and roof space (only applicable to rooms for residential purposes)**

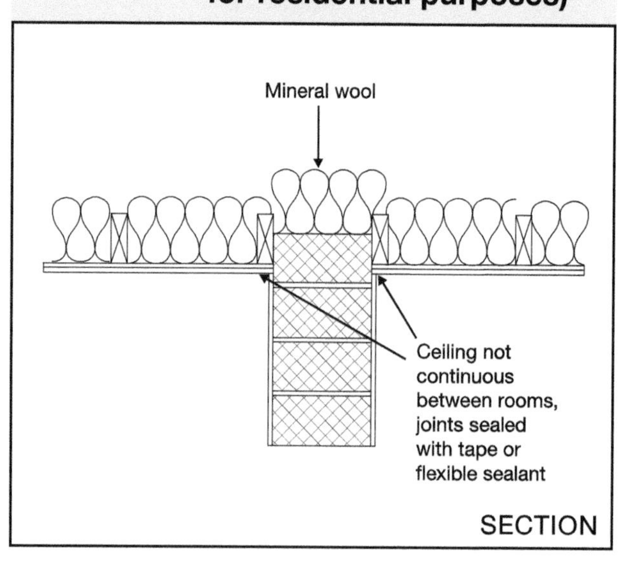

Room layout and building services design considerations

6.16 Internal noise levels are affected by room layout, building services and sound insulation.

6.17 The layout of rooms should be considered at the design stage to avoid placing noise sensitive rooms next to rooms in which noise is generated.

6.18 Additional guidance is provided in BS 8233:1999 Sound insulation and noise reduction for buildings. Code of practice and sound control for homes. See Annex D: References.

Section 7: Reverberation in the common internal parts of buildings containing flats or rooms for residential purposes

Introduction

7.1 This Section describes how to determine the amount of additional absorption to be used in corridors, hallways, stairwells and entrance halls that give access to flats and rooms for residential purposes.

7.2 For the purposes of this Section, a corridor or hallway is a space for which the ratio of the longest to the shortest floor dimension is greater than three.

7.3 For the purposes of this Section, an entrance hall is a space for which the ratio of the longest to the shortest floor dimension is three or less.

7.4 When an entrance hall, corridor, hallway or stairwell opens directly into another of these spaces, the guidance should be followed for each space individually.

7.5 The choice of absorptive material should meet the Requirements of Building Regulation Part B – Fire safety.

7.6 Two methods are described to satisfy Requirement E3, Method A and Method B.

7.7 **Method A:** Cover a specified area with an absorber of an appropriate class that has been rated according to BS EN ISO 11654:1997 Acoustics. Sound absorbers for use in buildings. Rating of sound absorption.

7.8 **Method B:** Determine the minimum amount of absorptive material using a calculation procedure in octave bands. Method B is intended only for corridors, hallways and entrance halls as it is not well suited to stairwells.

7.9 Where additional guidance is required, specialist advice should be sought at an early stage

Method A

7.10 For entrance halls, corridors or hallways, cover an area equal to or greater than the floor area, with a Class C absorber or better. It will normally be convenient to cover the ceiling area with the additional absorption.

7.11 For stairwells or a stair enclosure, calculate the combined area of the stair treads, the upper surface of the intermediate landings, the upper surface of the landings (excluding ground floor) and the ceiling area on the top floor. Either cover at least an area equal to this calculated area with a Class D absorber, or cover an area equal to at least 50% of this calculated area with a Class C absorber or better. The absorptive material should be equally distributed between all floor levels. It will normally be convenient to cover the underside of intermediate landings, the underside of the other landings, and the ceiling area on the top floor.

7.12 Method A can generally be satisfied by the use of proprietary acoustic ceilings. However, the absorptive material can be applied to any surface that faces into the space.

Method B

7.13 In comparison with Method A, Method B takes account of the existing absorption provided by all surfaces. In some cases, Method B should allow greater flexibility in meeting Requirement E3 and require less additional absorption than Method A.

7.14 For an absorptive material of surface area, S in m^2, and sound absorption coefficient, α the absorption area A is equal to the product of S and α.

7.15 The total absorption area, A_T, in square metres is defined as the hypothetical area of a totally absorbing surface, which if it were the only absorbing element in the space would give the same reverberation time as the space under consideration.

7.16 For n surfaces in a space, the total absorption area, A_T, can be found using the following equation.

$$A_T = \alpha_1 S_1 + \alpha_2 S_2 + ... + \alpha_n S_n$$

7.17 For entrance halls, provide a minimum of 0.20m^2 total absorption area per cubic metre of the volume. The additional absorptive material should be distributed over the available surfaces.

7.18 For corridors or hallways, provide a minimum of 0.25m^2 total absorption area per cubic metre of the volume. The additional absorptive material should be distributed over one or more of the surfaces.

7.19 Absorption areas should be calculated for each octave band. Requirement E3 will be satisfied when the appropriate amount of absorption area is provided for each octave band between 250Hz and 4000Hz inclusively.

7.20 Absorption coefficient data (to two decimal places) should be taken from the following:

- For specific products, use laboratory measurements of absorption coefficient data determined using BS EN 20354:1993 Acoustics. Measurement of sound absorption in a reverberation room. The measured third octave band data should be converted to practical sound absorption coefficient data, α_p in octave bands, according to BS EN ISO 11654:1997 Acoustics. Sound absorbers for use in buildings. Rating of sound absorption.

- For generic materials, use Table 7.1. This contains typical absorption coefficient data for common materials used in buildings. These data may be supplemented by published octave band data for other generic materials.

7.21 In Method B, each calculation step is to be rounded to two decimal places.

Table 7.1 **Absorption coefficient data for common materials in buildings**

Material	Sound absorption coefficient, α in octave frequency bands (Hz)				
	250	500	1000	2000	4000
Fair-faced concrete or plastered masonry	0.01	0.01	0.02	0.02	0.03
Fair-faced brick	0.02	0.03	0.04	0.05	0.07
Painted concrete block	0.05	0.06	0.07	0.09	0.08
Windows, glass façade	0.08	0.05	0.04	0.03	0.02
Doors (timber)	0.10	0.08	0.08	0.08	0.08
Glazed tile/marble	0.01	0.01	0.01	0.02	0.02
Hard floor coverings (e.g. lino, parquet) on concrete floor	0.03	0.04	0.05	0.05	0.06
Soft floor coverings (e.g. carpet) on concrete floor	0.03	0.06	0.15	0.30	0.40
Suspended plaster or plasterboard ceiling (with large air space behind)	0.15	0.10	0.05	0.05	0.05

Report format

7.22 Evidence that Requirement E3 has been satisfied should be presented, for example on a drawing or in a report, which should include:

1. A description of the enclosed space (entrance hall, corridor, stairwell etc.)
2. The approach used to satisfy Requirement E3, Method A or B.
   - With Method A, state the absorber class and the area to be covered.
   - With Method B, state the total absorption area of additional absorptive material used to satisfy the requirement.
3. Plans indicating the assignment of the absorptive material in the enclosed space.

Worked example

7.23 Example: Entrance hall

The entrance hall has dimensions 3.0m (width) x 4.0m (length) x 2.5m (height). The concrete floor is covered with carpet, the walls are painted concrete blocks and there are four timber doors (1.0m x 2.4m).

To satisfy Requirement E3, either use:

- Method A: Cover at least 3.0 x 4.0 = 12m^2 with a Class C absorber or better, or
- Method B: Provide a minimum of 0.2m^2 absorption area per cubic metre of the volume.

7.24 Method B is described in steps 1 to 8 in Table 7.2. In this example, the designer considers that covering the entire ceiling is a convenient way to provide the additional absorption. The aim of the calculation is to determine the absorption coefficient, $\alpha_{ceiling}$, needed for the entire ceiling.

7.25 In this example, the absorption coefficients from Method B indicate that a Class D absorber could be used to cover the ceiling. This can be compared against the slightly higher absorption requirement of Method A, which would have used a Class C absorber or better to cover the ceiling.

Table 7.2 **Example calculation for an entrance hall (Method B)**

Step 1 Calculate the surface area related to each absorptive material (i.e. for the floor, walls, doors and ceiling).

Surface	Surface finish	Area (m^2)
Floor	Carpet on concrete base	12.00
Doors	Timber	9.60
Walls (excluding door area)	Concrete block, painted	25.40
Ceiling	To be determined from this calculation	12.00

Step 2 Obtain values of absorption coefficients for the carpet, painted concrete block walls and the timber doors. In this case, the values are taken from Table 7.1.

		Absorption coefficient (α) in octave frequency bands				
Surface	**Area (m^2)**	**250Hz**	**500Hz**	**1000Hz**	**2000Hz**	**4000Hz**
Floor	12.00	0.03	0.06	0.15	0.30	0.40
Doors	9.60	0.10	0.08	0.08	0.08	0.08
Walls	25.40	0.05	0.06	0.07	0.09	0.08
Ceiling	12.00	To be determined from this calculation				

Step 3 Calculate the absorption area (m^2) related to each absorptive surface (i.e. for the floor, walls and doors) in octave frequency bands (Absorption area = surface area x absorption coefficient).

	Absorption area (m^2)				
Surface	**250Hz**	**500Hz**	**1000Hz**	**2000Hz**	**4000Hz**
Floor	0.36 (12.00 x 0.03)	0.72	1.80	3.60	4.80
Doors	0.96 (9.60 x 0.10)	0.77	0.77	0.77	0.77
Walls	1.27 (25.40 x 0.05)	1.52	1.78	2.29	2.03

Step 4 Calculate the sum of the absorption areas (m^2) obtained in Step 3.

	250Hz	**500Hz**	**1000Hz**	**2000Hz**	**4000Hz**
Total absorption area (m^2)	2.59 (0.36 + 0.96 + 1.27)	3.01	4.35	6.66	7.60

Step 5 Calculate the total absorption area (A_T) required for the entrance hall. The volume is $30m^3$ and therefore 0.2 x 30.0 = $6.0m^2$ of absorption area is required.

A_T (m^2)	6.00

Step 6 Calculate additional absorption area (A) to be provided by ceiling (m^2). If any values of minimum absorption area are negative, e.g. see 2000Hz and 4000Hz, then there is sufficient absorption from the other surfaces to meet the requirement without any additional absorption in this octave band (Additional absorption = A_T – total absorption area (from Step 5)).

	250Hz	**500Hz**	**1000Hz**	**2000Hz**	**4000Hz**
Additional absorption area (m^2)	3.41 (6.00 – 2.59)	2.99	1.65	-0.66	-1.60
				N.B. negative values indicate that no additional absorption is necessary.	

Step 7 Calculate required absorption coefficient (α) to be provided by ceiling (Required absorption coefficient α = Additional absorption area / area of ceiling).

	250Hz	**500Hz**	**1000Hz**	**2000Hz**	**4000Hz**
Required absorption coefficient, α	0.28 (3.41 ÷ 12.0)	0.25	0.14	Any value	Any value

Step 8 Identify a ceiling product from manufacturer's laboratory measurement data that provides absorption coefficients that exceed the values calculated in Step 7.

Section 8: Acoustic conditions in schools

8.1 In the Secretary of State's view the normal way of satisfying Requirement E4 will be to meet the values for sound insulation, reverberation time and internal ambient noise which are given in Building Bulletin 93. *Acoustic design of schools: performance standards* available on the internet at www.gov.uk.

Annex A: Method for calculating mass per unit area

A1 Wall mass

A1.1 Where a mass is specified it is expressed as mass per unit area in kilograms per square metre (kg/m²).

A1.2 The mass may be obtained from the manufacturer or it may be calculated by the method given in this annex. To calculate the mass per unit area of a masonry leaf use the formula below. This formula is not exact but is sufficient for this purpose.

A2 Formula for calculation of wall leaf mass per unit area

A2.1 Mass per unit area of a brick/block leaf = mass of co-ordinating area / co-ordinating area

$$= \frac{M_B + \rho_m (Td\,(L + H - d) + V)}{LH} \text{ kg/m}^2$$

where

- M_B is brick/block mass (kg) at appropriate moisture content
- ρ_m is density of mortar (kg/m³) at appropriate moisture content
- T is the brick/block thickness without surface finish (m)
- d is mortar thickness (m)
- L is co-ordinating length (m)
- H is co-ordinating height (m)
- V is volume of any frog/void filled with mortar (m³)

Note: This formula provides the mass per unit area of the block/brick construction without surface finish.

Note: See Diagram A.1 for block and mortar dimensions.

A2.2 When calculating the mass per unit area for bricks and blocks use the density at the appropriate moisture content from Table 3.2, CIBSE Guide A (1999).

A2.3 For cavity walls the mass per unit area of each leaf is calculated and added together.

A2.4 Where surface finishes are used the mass per unit area of the finish is added to the mass per unit area of the wall

Diagram A.1 **Block and mortar dimensions**

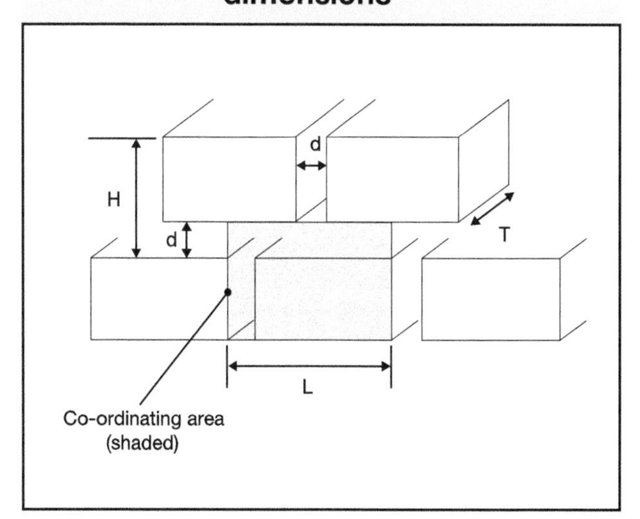

A3 Simplified equations

A3.1 Two examples are given (see Table A.1 and A.2) using the equation in A2.1. For each of these examples a simplified equation is obtained for that type of construction.

Table A.1 **Blocks laid flat**

Example of single leaf wall, blocks laid flat

- d = 0.010m
- T = 0.215m
- L = 0.450m
- H = 0.110m
- V = 0m³
- ρ_m = 1800kg/m³
- No surface finish

Mass per unit area = 20.2M_B + 43.0kg/m²

Substituting for M_B in this formula gives the following values:

Block mass, M_B (kg)	Mass per unit area (kg/m²)
6	164
8	205
10	245
12	285
14	326
16	366
18	407

Table A.2 Blocks laid on edge

Example of single leaf wall, blocks laid on edge

- $d = 0.010$m
- $T = 0.100$m
- $L = 0.450$m
- $H = 0.225$m
- $V = 0$m^3
- $\rho_m = 1800$kg/m^3
- No surface finish

Single leaf wall:
Mass per unit area = 9.9M_B + 11.8kg/m^2

Cavity wall:
Mass per unit area = 19.8M_B + 23.6kg/m^2

Substituting for M_B in this formula gives the following values:

Block mass, M_B (kg)	Mass per unit area (kg/m^2)	
	Single leaf	Cavity
6	71	142
8	91	182
10	111	222
12	131	261
14	150	301
16	170	340
18	190	380

A4 Mass per unit area of surface finishes

A4.1 The mass per unit area of surface finishes should be obtained from manufacturer's data.

A5 Mass per unit area of floors

A5.1 The mass of a solid and homogeneous floor (without hollows, beams or ribs) can be calculated from:

$$M_F = \rho_c \times T$$

where,

M_F is mass per unit area of floor (kg/m^2)

ρ_c is density of concrete (kg/m^3)

T is thickness of floor (m)

A5.2 The mass of a beam and block floor can be calculated from:

$$M_F = (M_{beam,1m} + M_{block,1m}) / L_B$$

where

M_F is mass per unit area of floor (kg/m^2)

$M_{beam,1m}$ is the mass of a 1m length of beam (kg)

$M_{block,1m}$ is the mass of a 1m length of blocks (kg)

L_B is the distance between the beam centre lines, i.e. the repetition interval (m)

Note: See Diagram A.2 for beam and block floor dimensions.

Diagram A.2 Beam and block floor dimensions

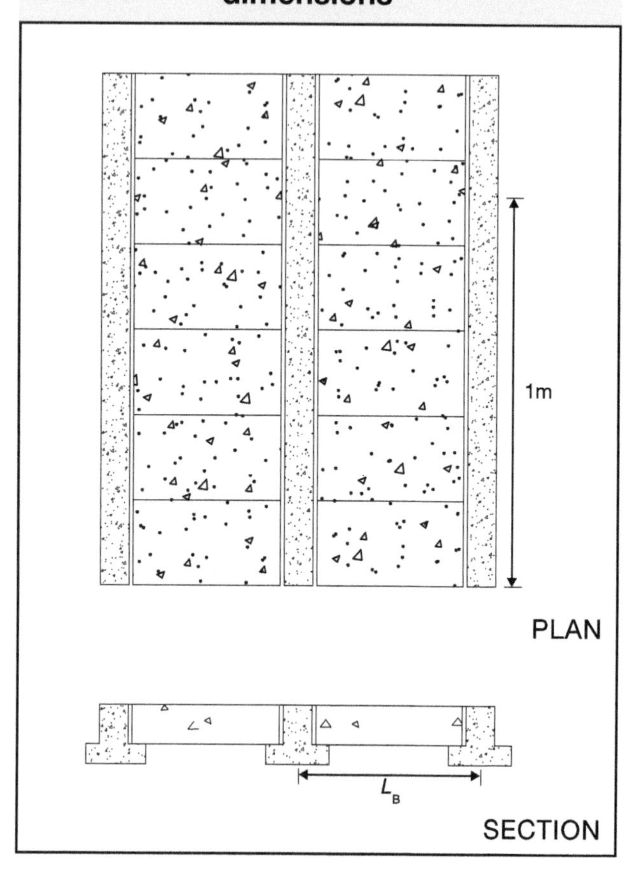

A5.3 For other floor types (including floors with variable thickness), seek advice from the manufacturer on mass per unit area and performance.

Annex B: Procedures for sound insulation testing

B1 Introduction

B1.1 Section B.2 of this Annex describes the sound insulation testing procedure approved by the Secretary of State for the purposes of Regulation 41(2)(a) of the Building Regulations and Regulation 20(1) of the Approved Inspectors Regulations. The approved procedure is that set out in Section B.2 and the Standards referred to in that Section.

B1.2 Section B.3 of this Annex provides guidance on laboratory testing in connection with achieving compliance with Requirement E2 in Schedule 1 to the Building Regulations, and in connection with evaluation of components to be used in constructions subject to Requirement E1.

B1.3 Section B.4 of this Annex gives guidance on test reports.

B1.4 The person carrying out the building work should arrange for sound insulation testing to be carried out by a test body with appropriate third party accreditation. Test bodies conducting testing should preferably have UKAS accreditation (or a European equivalent) for field measurements. The ODPM also regards members of the ANC Registration Scheme as suitably qualified to carry out pre-completion testing. The measurement instrumentation used should have a valid, traceable certificate of calibration, and should have been tested within the past two years.

B2 Field measurement of sound insulation of separating walls and floors for the purposes of Regulation 41 and Regulation 20(1) and (5)

Introduction

B2.1 Sound insulation testing for the purposes of Regulation 41 of the Building Regulations and Regulation 20(1) and (5) of the Approved Inspectors Regulations 2010, must be done in accordance with: BS EN ISO 140-4:1998; BS EN ISO 140- 7:1998; BS EN ISO 717-1:1997; BS EN ISO 717-2:1997; BS EN 20354:1993. When calculating sound insulation test results, no rounding should occur in any calculation until required by the relevant Standards, the BS EN ISO 140 series and the BS EN ISO 717 series.

Airborne sound insulation of a separating wall or floor

B2.2 The airborne sound insulation of a separating wall or floor should be measured in accordance with BS EN ISO 140-4:1998. All measurements and calculations should be carried out using one-third octave frequency bands. Performance should be rated in terms of the weighted standardised level difference, $D_{nT,w}$, and spectrum adaptation term, C_{tr}, in accordance with BS EN ISO 717-1:1997.

Measurements using a single sound source

B2.3 For each source position, the average sound pressure level in the source and receiving rooms is measured in one-third octave bands using either fixed microphone positions (and averaging these values on an energy basis) or a moving microphone.

B2.4 For the source room measurements, the difference between the average sound pressure levels in adjacent one-third octave bands should be no more than 6dB. If this condition is not met, the source spectrum should be adjusted and the source room measurement repeated. If the condition is met, the average sound pressure level in the receiving room, and hence a level difference, should be determined.

B2.5 It is essential that all measurements made in the source and receiving rooms to determine a level difference should be made without moving the sound source or changing the output level of the sound source, once its spectrum has been correctly adjusted (where necessary).

B2.6 The sound source should now be moved to the next position in the source room and the above procedure repeated to determine another level difference. At least two positions should be used for the source. The level differences obtained from each source position should be arithmetically averaged to determine the level difference, *D* as defined in BS EN ISO 140-4:1998.

Measurements using multiple sound sources operating simultaneously

B2.7 For multiple sound sources operating simultaneously, the average sound pressure level in the source and receiving rooms is measured in one-third octave bands using either fixed microphone positions (and averaging these values on an energy basis) or a moving microphone.

B2.8 For the source room measurements, the difference between the average sound pressure levels in adjacent one-third octave bands should be no more than 6dB. If this condition is not met, the source spectrum should be adjusted and the source room measurement repeated. If the condition is met, determine the average level in the receiving room, and hence the level difference, *D* as defined in BS EN ISO 140-4:1998.

Impact sound transmission of a separating floor

B2.9 The impact sound transmission of a separating floor should be measured in accordance with BS EN ISO 140-7:1998. All measurements and calculations should be carried out using one-third-octave frequency bands. Performance should be rated in terms of the weighted standardised impact sound pressure level, $L'_{nT,w}$ in accordance with BS EN ISO 717-2:1997.

Measurement of reverberation time

B2.10 BS EN ISO 140-4:1998 and BS EN ISO 140-7:1998 refer to the ISO 354 (BS EN 20354:1993) method for measuring reverberation time. However, for the approved procedure, the guidance in BS EN ISO 140-7:1998 relating to the source and microphone positions, and the number of decay measurements required, should be followed.

Room requirements

B2.11 Section 1 gives guidance on the room types that should be used for testing. These rooms should have volumes of at least 25m^3. If this is not possible then the volumes of the rooms used for testing should be reported.

Tests between rooms

B2.12 Tests should be conducted in completed but unfurnished rooms or available spaces in the case of properties sold before fitting out; see Section 1.

B2.13 Impact sound insulation tests should be conducted on a floor without a soft covering (e.g. carpet, foam backed vinyl) except in the case of (a) separating floor type 1, as described in this Approved Document, or (b) a concrete structural floor base which has a soft covering as an integral part of the floor.

B2.14 If a soft covering has been installed on any other type of floor, it should be taken up. If that is not possible, at least half of the floor should be exposed and the tapping machine should be placed only on the exposed part of the floor.

B2.15 When measuring airborne sound insulation between a pair of rooms of unequal volume, the sound source should be in the larger room.

B2.16 Doors and windows should be closed.

B2.17 Kitchen units, cupboards etc. on all walls should have their doors open and be unfilled.

Measurement precision

B2.18 Sound pressure levels should be measured to 0.1dB precision.

B2.19 Reverberation times should be measured to 0.01s precision.

Measurements using a moving microphone

B2.20 At least two positions should be used.

B2.21 For measurements of reverberation time, discrete positions should be used rather than a moving microphone.

B3 Laboratory measurements

Introduction

B3.1 Pre-completion testing for the purposes of Regulation 41 and Regulation 20(1) and (5) involves field testing on separating walls and floors (see Section 1 and Annex B: B2). However, there are applications for laboratory tests to determine the performance of: floor coverings; floating floors; wall ties; resilient layers; internal walls and floors; and flanking laboratory tests to indicate the performance of novel constructions.

B3.2 When calculating sound insulation test results, no rounding should occur in any calculation until required by the relevant Standards, i.e. the BS EN ISO 140 series and the BS EN ISO 717 series.

Tests on floor coverings and floating floors

B3.3 Floor coverings and floating floors should be tested in accordance with BS EN ISO 140-8:1998 and rated in accordance with BS EN ISO 717-2:1997. The test floor should have a thickness of 140mm.

B3.4 It should be noted that text has been omitted from BS EN ISO 140-8:1998. For the purposes of this Approved Document, Section 6.2.1 of BS EN ISO 140-8:1998 should be disregarded, and Section 5.3.3 of BS EN ISO 140-7:1998, respectively, referred to instead.

B3.5 BS EN ISO 140-8:1998 refers to the ISO 354 (BS EN 20354:1993) method for measuring reverberation time, but the guidance in BS EN ISO 140-8:1998 relating to the source and microphone positions, and the number of decay measurements required, should be followed.

B3.6 When assessing category II test specimens (as defined in BS EN ISO 140-8:1998) for use with separating floor type 2, the performance value (ΔL_w) should be achieved when the floating floor is both loaded and unloaded. The loaded measurements should use a uniformly distributed load of 20–25kg/m^2 with at least one weight per square metre of the flooring area, as described in BS EN ISO 140-8:1998.

Dynamic stiffness of resilient layers

B3.7 Dynamic stiffness of resilient layers should be measured in accordance with BS EN 29052-1:1992. The test method using sinusoidal signals should be used. No pre-compression should be applied to the test specimens before the measurements.

Dynamic stiffness of wall ties

B3.8 Dynamic stiffness of wall ties should be measured in accordance with BRE Information Paper IP 3/01.

Airborne sound insulation of internal wall and floor elements

B3.9 The airborne sound insulation of internal wall or floor elements in a laboratory should be measured in accordance with BS EN ISO 140-3:1995, and the performance rated in accordance with BS EN ISO 717-1:1997 to determine the weighted sound reduction index, R_w.

Measurements in a flanking laboratory

B3.10 Tests of sound transmission in a flanking laboratory include both direct and flanking paths, and are a useful means of assessing the likely field performance of novel constructions.

B3.11 It is not possible to demonstrate compliance with Requirement E1 using test results from a flanking laboratory.

Flanking laboratory: design

B3.12 Construction details of a suitable laboratory can be obtained from the Acoustics Centre, BRE, Garston, Watford WD25 9XX.

Note: A CEN standard for the laboratory measurement of flanking transmission between adjoining rooms is currently under development.

Flanking laboratory: indicative airborne sound insulation values

B3.13 When a test construction has airborne sound insulation of at least 49dB $D_{nT,w} + C_{tr}$ when measured in a flanking laboratory using the procedure given in Annex B: B2, this can be taken as indicative that the same construction (i.e. identical in all significant details) may achieve at least 45dB $D_{nT,w} + C_{tr}$ when built in the field. See paragraph B3.11.

Flanking laboratory: indicative impact sound insulation values

B3.14 When a test construction has impact sound insulation no more than 58dB $L'_{nT,w}$ when measured in a flanking laboratory using the procedure given in Annex B: B2, this can be taken as indicative that the same construction (i.e. identical in all significant details) may achieve no more than 62dB $L'_{nT,w}$ when built in the field. See paragraph B3.11.

B4 Information to be included in test reports

Field test reports

B4.1 Paragraph 1.41 of this Approved Document sets out the manner of recording the results of testing done for the purposes of Regulation 41 or Regulation 20(1) and (5), approved by the Secretary of State under those Regulations.

Although not required, it may be useful to have a description of the building including:

1. sketches showing the layout and dimensions of rooms tested;
2. description of separating walls, external walls, separating floors, and internal walls and floors including details of materials used for their construction and finishes;
3. mass per unit area in kg/m^2 of separating walls, external walls, separating floors, and internal walls and floors;
4. dimensions of any step and/or stagger between rooms tested;
5. dimensions and position of any windows or doors in external walls.

Laboratory test reports for internal walls and floors

B4.2 Test reports should include the following information.

1. Organisation conducting test, including:
   a. name and address;
   b. third party accreditation number (e.g. UKAS or European equivalent);
   c. Name(s) of person(s) in charge of test.
2. Name(s) of client(s).
3. Date of test.
4. Brief details of test, including:
   a. equipment;
   b. test procedures.
5. Full details of the construction under test and the mounting conditions.
6. Results of test shown in tabular and graphical form for one-third octave bands according to the relevant part of the BS EN ISO 140 series and BS EN ISO 717 series, including:
   a. single-number quantity and the spectrum adaptation terms;
   b. data from which the single-number quantity is calculated.

Annex C: Glossary

The definitions given below are for the purposes of this document only, and are not intended to be rigorous. Fuller definitions of the various acoustical terms are to be found in the relevant British Standards listed in Annex D.

Absorption
Conversion of sound energy to heat, often by the use of a porous material.

Absorption coefficient
A quantity characterising the effectiveness of a sound absorbing surface. The proportion of sound energy absorbed is given as a number between zero (for a fully reflective surface) and one (for a fully absorptive surface). Note that sound absorption coefficients determined from laboratory measurements may have values slightly larger than one. See BS EN 20354:1993.

Absorptive material
Material that absorbs sound energy.

Airborne sound
Sound propagating through the air.

Airborne sound insulation
Sound insulation that reduces transmission of airborne sound between buildings or parts of buildings.

Air path
A direct or indirect air passage from one side of a structure to the other.

Caulking
Process of sealing joints.

Cavity stop
A proprietary product or material such as mineral wool used to close the gap in a cavity wall.

$\boldsymbol{C}_{tr}$
The correction to a sound insulation quantity (such as $D_{nT,w}$) to take account of a specific sound spectrum. See BS EN ISO 717-1:1997.

dB
(See decibel)

Decibel (dB)
The unit used for many acoustic quantities to indicate the level with respect to a reference level.

Density
Mass per unit volume, expressed in kilograms per cubic metre (kg/m^3).

Direct transmission
The process in which sound that is incident on one side of a building element is radiated by the other side.

$\boldsymbol{D}_{nT}$

The difference in sound level between a pair of rooms, in a stated frequency band, corrected for the reverberation time. See BS EN ISO 140-4:1998.

$\boldsymbol{D}_{nT,w}$
A single-number quantity which characterises the airborne sound insulation between rooms. See BS EN ISO 717-1:1997.

$\boldsymbol{D}_{nT,w} + \boldsymbol{C}_{tr}$
A single-number quantity which characterises the airborne sound insulation between rooms using noise spectrum no. 2 as defined in BS EN ISO 717-1:1997. See BS EN ISO 717-1:1997.

Dynamic stiffness
A parameter used to describe the ability of a resilient material or wall tie to transmit vibration. Specimens with high dynamic stiffness (dynamically 'stiff') transmit more vibration than specimens with low dynamic stiffness (dynamically 'soft'). See BS EN 29052-1:1992 for resilient materials. See BRE Information Paper IP 3/01 for wall ties.

Flanking element
Any building element that contributes to sound transmission between rooms in a building that is not a separating floor or separating wall.

Flanking transmission
Sound transmitted between rooms via flanking elements instead of directly through separating elements or along any path other than the direct path.

Floating floor
A floating floor consists of a floating layer and resilient layer (see also resilient layer and floating layer).

Floating layer
A surface layer that rests on a resilient layer and is therefore isolated from the base floor and the surrounding walls (see also resilient layer).

Framed wall
A partition consisting of board or boards connected to both sides of a wood or metal frame.

Frequency
The number of pressure variations (or cycles) per second that gives a sound its distinctive tone. The unit of frequency is the Hertz (Hz).

Frequency band
A continuous range of frequencies between stated upper and lower limits (see also octave band and one-third octave band).

Hertz (Hz)
The unit of the frequency of a sound (formerly called cycles per second).

Impact sound
Sound resulting from direct impact on a building element.

Impact sound insulation
Sound insulation which reduces impact sound transmission from direct impacts such as footsteps on a building element.

Independent ceiling
A ceiling which is fixed independently of a separating floor or an internal floor (see separating floor and internal floor).

Internal floor
Any floor that is not a separating floor (see separating floor).

Intermediate landing
A landing between two floors (see also landing).

Internal wall
Any wall that does not have a separating function.

Isolation
The absence of rigid connections between two or more parts of a structure.

Landing
A platform or part of floor structure at the end of a flight of stairs or ramp.

L'_{nT}
The impact sound pressure level in a stated frequency band, corrected for the reverberation time. See BS EN ISO 140-7:1998.

$L'_{nT,w}$
A single-number quantity used to characterise the impact sound insulation of floors. See BS EN ISO 717-2:1997.

Mass per unit area
Mass per unit area is expressed in terms of kilograms per square metre (kg/m^2).

Noise
Noise is unwanted sound.

Octave band
A frequency band in which the upper limit of the band is twice the frequency of the lower limit.

One-third octave band
A frequency band in which the upper limit of the band is $2^{1/3}$ times the frequency of the lower limit.

R_w
A single-number quantity which characterises the airborne sound insulation of a material or building element in the laboratory. See BS EN ISO 717-1:1997.

Resilient layer
A layer that isolates a floating layer from a base floor and surrounding walls.

Reverberation
The persistence of sound in a space after a sound source has been stopped.

Reverberation time
The time, in seconds, taken for the sound to decay by 60dB after a sound source has been stopped.

Separating floor
Floor that separates flats or rooms for residential purposes.

Separating wall
Wall that separates adjoining dwelling-houses, flats or rooms for residential purposes.

Sound pressure level
A quantity related to the physical intensity of a sound.

Sound reduction index (*R*)
A quantity, measured in a laboratory, which characterises the sound insulating properties of a material or building element in a stated frequency band. See BS EN ISO 140-3:1995.

Spectrum
The composition of a particular sound in terms of separate frequency bands.

Structure-borne sound
Sound which is carried via the structure of a building.

UKAS
United Kingdom Accreditation Service.

ΔL_w
The measured improvement of impact sound insulation resulting from the installation of a floor covering or floating floor on a test floor in a laboratory. See BS EN ISO 717-2:1997.

Annex D: References

D1 STANDARDS

BS Series

BS 1243:1978
Metal ties for cavity wall construction. AMD 3651 1981, AMD 4024 1982.
(Withdrawn and superseded by BS EN 845-1:2000 Specification for ancillary components for masonry. Ties tension straps, hangers and brackets. AMD 14736 2003.)

BS 1289-1:1986
Flue blocks and masonry terminals for gas appliances. Specification for precast concrete flue blocks and terminals. AMD 9853 1998.
(Withdrawn and superseded by BS EN 1858:2003 Chimneys. Components. Concrete flue blocks.)

BS 5628-3:2001
Code of practice for use of masonry. Materials and components, design and workmanship.

BS 8233:1999
Sound Insulation and noise reduction for buildings. Code of practice.

BS EN Series

BS EN 20354:1993
Acoustics. Measurement of sound absorption in a reverberation room. AMD 7781 1993, AMD 9974 1998.
(Withdrawn and superseded by BS EN ISO 354:2003 Acoustics. Measurement of Sound absorption in a reverberation room.
AMD 14766 2003.)

BS EN 29052-1:1992
Acoustics. Method for the determination of dynamic stiffness. Materials used under floating floors in dwellings.

BS EN ISO Series

BS EN ISO 140-3:1995
Acoustics. Measurement of sound insulation in buildings and of building elements. Laboratory measurement of airborne sound insulation of building elements. AMD 15277 2005. (Also known as BS 2750-3:1995.)

BS EN ISO 140-4:1998
Acoustics. Measurement of sound insulation in buildings and of building elements. Field measurements of airborne sound insulation between rooms.

BS EN ISO 140-6:1998
Acoustics. Measurement of sound insulation in buildings and of building elements. Laboratory measurements of impact sound insulation of floors.

BS EN ISO 140-7:1998
Acoustics. Measurement of sound insulation in buildings and of building elements. Field measurements of impact sound insulation of floors.

BS EN ISO 140-8:1998
Acoustics. Measurement of sound insulation in buildings and of building elements. Laboratory measurements of the reduction of transmitted impact noise by floor coverings on a heavyweight standard floor.

BS EN ISO 717-1:1997
Acoustics. Rating of sound insulation in buildings and of building elements. Airborne sound insulation.

BS EN ISO 717-2:1997
Acoustics. Rating of sound insulation in buildings and of building elements. Impact sound insulation.

BS EN ISO 11654:1997
Acoustics. Sound absorbers for use in buildings. Rating of sound absorption.

D2 GUIDANCE

BRE

Information Paper IP 3/01 *Dynamic stiffness of wall ties used in masonry cavity walls: measurement procedure*, 2001. ISBN 1 86081 461 1

Information Paper IP 4/01 *Reducing impact and structure borne sound in buildings*, 2001. ISBN 1 86081 462 X

Information Paper IP 14/02 *Dealing with poor sound insulation between new dwellings*, 2002. ISBN 1 86081 549 0

Report BR 262 *Thermal insulation: avoiding risks*, 2002, ISBN 1 86081 515 4

Report BR 238 *Sound Control for Homes*, 1993. ISBN 0 85125 559 0. Joint publication with CIRIA Report 127 ISBN 0 86017 362 3. Note: some of the information within this document has been superseded.

CIBSE

Guide A *Environmental design*, 6th edition, 1999. ISBN 0 90095 396 9

Department for Education and Skills (DfES)

Building Bulletin 93 *Acoustic design of schools: performance standards,* 2015. www.gov.uk.

D3 LEGISLATION

HSE

L24 *The Workplace (Health, Safety and Welfare) Regulations 1992. Approved Code of Practice and Guidance*, 1992. ISBN 0 71760 413 6

Building Act 1984, Chapter 55.

Construction Products Regulations 1991, SI 1991/1620.

Construction Products (Amendment) Regulations 1994, SI 1994/3051.

Construction Products Directive (89/106/EEC).

The Gas Safety (Installation and Use) Regulations 1998, SI 1998/2451.

CE marking Directive (93/68/EEC).

Annex E: Design details approved by Robust Details Ltd

Robust Details Ltd is a non-profit distributing company, limited by guarantee, set up by the house-building industry. Its objectives are broadly to identify, arrange testing and, if satisfied, approve and publish design details that, if correctly implemented in separating structures, should achieve compliance with Requirement E1.
It also carries out checks on the performance achieved in practice.

The robust design details are available in a handbook, which may be purchased from Robust Details Ltd. The company can be contacted at: PO Box 7289, Milton Keynes, Bucks, MK14 6ZQ; telephone 0870 240 8210; fax 0870 240 8203; e-mail administration@robustdetails.com; website www.robustdetails.com

Although the design details are in the public domain, their use in building work is not authorised unless the builder has registered the particular use of the relevant design detail or details with Robust Details Ltd and obtained a unique number or numbers from the company. Each unique number identifies a house or flat in which one or more of the design details are being used.

The system of unique numbers makes possible an essential part of Robust Details Ltd's procedures for ensuring that design details it has approved deliver reasonable sound insulation performance in practice. Robust Details Ltd carries out a programme of checks on a proportion of cases where approved design details are used.

Under Regulation 41(4) of the Building Regulations 2010 and Regulation 20(1) of the Building (Approved Inspector, etc.) Regulations 2010, the requirement for appropriate sound insulation testing imposed by Regulations 41 and 20(1) does not apply to parts of the building which would otherwise be subject to the testing requirement where all the following apply:

a. the building work consists of the erection of a new dwelling-house (i.e. a semi-detached or terraced house) or a building containing flats;

b. the person carrying out the building work notifies the building control body before the start of building work on site that, in a specified part or parts of the building, he is using one or more specified design details from those approved by Robust Details Ltd. In a case where building control is being carried out by the local authority, the notification must be given not later than the date on which notice of commencement of construction is given under Regulation 16(1) of the Building Regulations 2010;

c. the notification specifies the unique number or numbers issued by Robust Details Ltd in respect of the specified use of the design detail or details;

d. the building work carried out in respect of the part or parts of the building identified in the notification is in accordance with the design detail or details specified in the notification.

If the notification is late, or if it does not specify the relevant part or parts, the design detail or details in question and the unique number or numbers, the part or parts of the building in question *are subject to sound insulation testing under Regulation 41 or 20(1) and (5) in the usual way.*

If the notification is itself valid but the work is not carried out in accordance with the design detail or details, the relevant separating structures become subject to sound insulation testing under Regulation 41 or 20(1) and (5). It would be open to the builder to take remedial action such that the building control body was satisfied that the work had been brought into compliance with the specified detail or details. *With that exception, testing under Regulation 41 or 20(1) and (5) would be needed on all structures that have been subject to a valid notification under Regulation 41(4) or 20(1) and (5) but which in the opinion of the building control body have not then been constructed in accordance with the specified detail or details.*

It should be noted that the compliance of work with a robust detail, in circumstances where the correct procedures have been followed to attract exemption from PCT, is not a 'deemed to satisfy' condition. The underlying requirement remains to achieve compliance with Part E1. The guidance in Approved Document E is that compliance will usually be established by the measured performance of the structure. Therefore it would be open to anyone, e.g. a homeowner, who considered that a party structure does not comply with Part E1, to seek to establish that by the carrying out of tests. It would **not** be a defence for the builder to show that he had correctly carried out a design detail approved by Robust Details Ltd, if the structure's measured performance were shown not to meet the performance standards in Approved Document E.